Land-Use and Land-Cover Changes

ATMOSPHERIC AND OCEANOGRAPHIC SCIENCES LIBRARY

VOLUME 44

For further volumes:
http://www.springer.com/series/5669

Nicole Mölders

Land-Use and Land-Cover Changes

Impact on Climate and Air Quality

Nicole Mölders
Department of Atmospheric Sciences
College of Natural Sciences
and Mathematics
Geophysical Institute
University of Alaska Fairbanks
903 Koyukuk Drive
Fairbanks, AK 99775
USA
molders@gi.alaska.edu
cmoelders@alaska.edu

ISSN 1383-8601
ISBN 978-94-007-1526-4 e-ISBN 978-94-007-1527-1
DOI 10.1007/978-94-007-1527-1
Springer Dordrecht Heidelberg London New York

Library of Congress Control Number: 2011936018

Printed on acid-free paper

Springer is part of Springer Science+Business Media (www.springer.com)

Acknowledgments

I would like to thank the most my father – Hermann W. Mölders – for evoking my interest in land-cover-related weather and climate change. He brought the climate–land-cover relations of the Sahel drought and potential climate impacts of irrigation to my attention when I was a teenager.

I also would like to express my thanks to Prof. Dr. Hantel and Prof. Dr. Kramm who encouraged me to write this book. I wish to thank my colleague Prof. Dr. Kramm and my graduate student Huy N.Q. Tran for fruitful discussions and helpful comments during the process of writing this book.

Thanks go also to my former mentors Prof. Dr. Raschke, Prof. Dr. Ramond, Prof. Dr. Pointin, Prof. Dr. Ebel, Dr. Laube, Prof. C. Walcek, Prof. Dr. Tetzlaff, and Dr. Raabe who helped me to become the scientist I am today.

Contents

Chapter 1

Introduction

Land-cover and land-cover changes (LCC) affect the moisture and temperature states as well as the composition of the atmospheric boundary layer (ABL). The often-unpleasant conditions in cities in summer are one of the most obvious examples of how land-use, that is the realization of the actual land-cover affects local weather and climate and, indirectly, air quality.

Land-cover changes can range from the modification of the landscape character without affecting the existing overall classifications to the extreme case, where one land-cover type completely replaces another. The latter is often referred to as land conversion (LC). Land-cover modification refers to anthropogenic (e.g., deforestation for agricultural expansion) or naturally caused (e.g., flooding, wildfire, disease, epidemics) LCC. In both cases, the replaced spatial entities can fall into a different land-cover category. In the first case, the replaced spatial entities may be just an alteration of extent, shape, and/or shift in location of the previous land-cover/use, or fragments or coalescence of a land-cover class. The location, temporal and spatial scales of various LCC differ among each other depending on the causes for the LCC.

Since humankind, the Earth's landscapes experienced naturally caused and anthropogenic LCC as well as local LC. Since on a small scale, in the sense of the micro-scale/meso-γ-scale classification by Orlanski (1975) (Fig. 1.1), all LCC can be viewed as LC, the distinction is not made any further throughout this book. Furthermore, no distinction between land-cover and actual land-use, and LCC or land-use changes will be made except if it serves for better understanding.

N. Mölders, *Land-Use and Land-Cover Changes*, Atmospheric and Oceanographic Sciences Library 44, DOI 10.1007/978-94-007-1527-1_1,

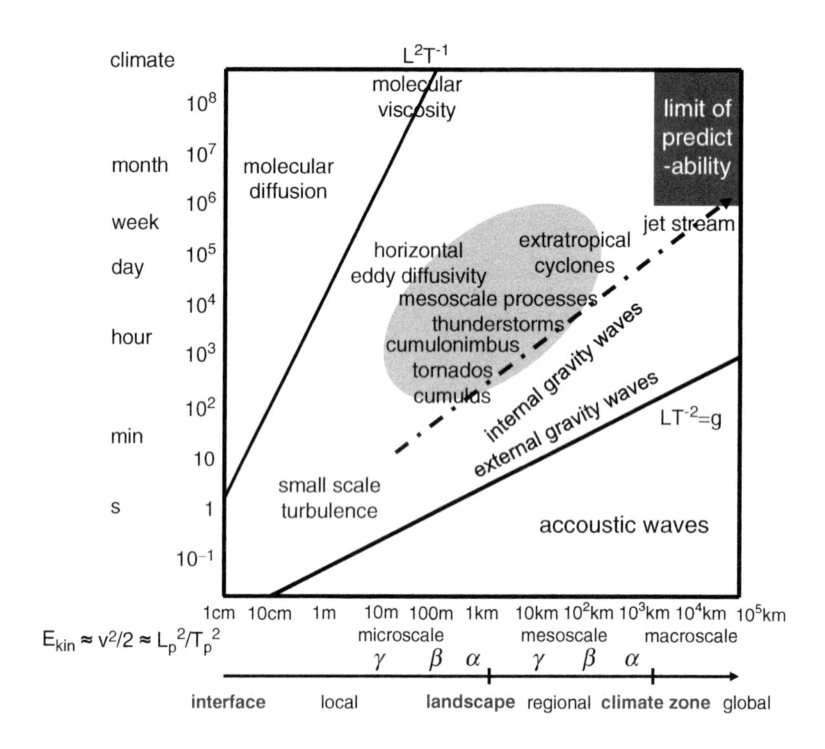

Figure 1.1. Definitions and different processes associated with characteristic temporal and spatial scales (Adapted from Orlanski (1975))

1.1 Natural Land-Cover Changes

Land-cover not only affects the local atmospheric conditions, but also is itself affected by local weather and depends on local climate. Consequently, land-cover may change naturally in response to extreme weather events that cause wildfires or lasting flooding, or multiple-year droughts. Land-cover changes in response to climate change include sea-level rise, invasion of non-native plants and shifts in ecosystem boundaries. Other driving forces for natural LCC are earthquake or otherwise induced landslides, and volcanic eruptions. In a broader sense of surface changes, we may consider break-off of ice shelves, glacier retreat, permafrost retreat and/or warming as near-surface responses to climate change.

Natural LCC may occur gradually or abrupt. Examples for gradually occurring LLC are desertification, non-native invasion of plants, shifts in ecosystem boundaries, or wildfire-succession landscapes. The Younger Dryas provides an example for LCC induced by climate change. The Younger Dryas was an abrupt cold event in the Northern Hemisphere during the last deglaciation between ~12,800 and 11,500 BP. Once the ice had receded far enough, the continental runoff that formerly discharged into the Gulf of Mexico via the Mississippi River partly discharged to the St. Lawrence River. The increased freshwater inflow into the North Atlantic affected the thermohaline circulation and reduced the poleward ocean-heat transport. The resulting subsequently altered ocean-surface conditions led to more sea-ice and a decrease of about 5 K in near-surface temperatures. In response to the cooler "climate," glacial tundra replaced the forests in Scandinavia (Munro 2003).

Examples of rather abrupt surface changes are associated with glacier retreat and often-related ice-shelf break-off. Glaciers and shelves calf naturally; however, the frequency and amount may be affected by climate change. Time-series of the extent of nine Antarctic Peninsula ice shelves, for instance, show that the five northern shelves have retreated radically (Vaughan and Doake 1996) since the third International Polar Year (1957–1958). Between 1966 and 1989, the Antarctic Wordie Ice Shelf decreased by $\approx$1,300 km^2 (Rott et al. 1996). In January 1995, 4,200 km^2 of the northern Antarctic Larsen Ice Shelf broke off after a time of steady retreat in response to regional warming. Satellite radar images of the Larsen Ice Shelf and neighboring glaciers showed further retreat of several kilometers inland of the previous grounding line after the 1995 collapse (Rott et al. 2002).

Ice-shelf retreat/collapse is not unique to Antarctica. Between 2000 and 2002 the largest Arctic ice shelf, the Ward Hunt Ice Shelf in Canada, broke off; in consequence, an epishelf lake – a rare ecosystem type – was lost (Mueller et al. 2003).

Some natural LCC may be self re-enforcing, while most natural LCC are temporary and self-correcting. Prominent examples for self re-enforcing LCC are snow-covered land, ice caps and glaciers. A change in the extent of these land-surface types increases the region's albedo, which may lead to cooling. In the relatively cooler atmosphere, the partitioning of precipitation shifts further in favor of snow, which may further extend the area of snow-cover and glacier. Over oceans, lower temperature may enhance the formation of sea ice. This positive climate-feedback process is known as the snow-albedo or albedo-temperature feedback.

Fires are another example for self-correcting LCC despite the fact that they work through a biological rather than (geo)physical mechanism. Fires are the most important natural disturbance agent that leads annually to over $3 \cdot 10^6$ km^2 of LCC worldwide. Since millennia, lightning-caused wildfires have initiated a natural gradual changing of landscapes with fire-adapted ecosystems. In such ecosystems, regrowth requires months to decades or longer depending on fuel type. All fire-succession landscapes show little to no biomass growth in the year of the fire. Grass and herbs start growing on the burned area in the first years after the fire followed by shrubs and, in naturally forest-dominated regions, trees (Mölders and Kramm 2007).

In case of self-correction of land-cover after fires, the time for self-correction depends on the climate zone and original land-cover and has implications for the frequency of wildfire occurrence. Boreal forest, for instance, requires about 80 years for recovery. Burned tundra recovers 50–100% after 5–6 years, while concurrently the thawing of soil stabilizes (Racine et al. 1987). Since grass and shrubs regenerate fast, landscapes with high presence of fine fuel may experience fires with larger horizontal extent and in shorter intervals between fires than landscapes dominated by woody fuel.

Economic and political changes may affect natural LCC, too. For instance, fire-management policies and changes thereof affect the annual area burned and the location where LCC due to wildfires occur. Invasive fire-prone species, fire suppression, prescribed burnings and related fuel changes, clearance, and intensive grazing increase fire susceptibility. Economic changes affecting subsistence and agricultural economies, tourism and recreation have led to an increase in population at the wildland-rural border. Consequently, the extension of areas with different fire-management degrees has changed over time. Such changes may have consequences for the disturbance regimes, ecosystem dynamics, biodiversity, carbon storage, emission of trace gases and particles, and atmospheric composition.

1.2 Anthropogenic Land-Cover Changes

Anthropogenic LCC or LC occur because of changing management practices or altered purpose (e.g., logging, ranching, cropping, construction, water storage) of land-use (e.g., forest, grassland, cropland, urban land, artificial lakes). Typically, anthropogenic LCC will persist for long time if the purpose of the LCC is a new land use. Anthropogenic LCC occur for food production (agriculture, grazing), biofuel production, establishing and/or growth of settlements and industrial areas, forest harvest and managing, and creation of water reservoirs, just to mention a few purposes. The reasons for LCC and their extension differ regionally and temporally. Since 1850, about $4.7 \cdot 10^6\,\mathrm{km}^2$ of savannahs, grasslands, and steppes and $6 \cdot 10^6\,\mathrm{km}^2$ of woodlands and forests were converted to croplands (Sivakumar 2007).

Anthropogenic LCC may occur in response to natural LCC and/or climate changes. The Younger Dryas – despite its initial LCC were caused naturally – provides also an example for climate-induced anthropogenic LCC. The cold and dry conditions of the Younger Dryas reduced the availability of food and forced the population into a more mobile subsistence life style. The continuing climatic deterioration may have led to cereal cultivation at the end of the Younger Dryas (Munro 2003). Concurrently, settlements started along rivers.

Today's landscapes developed from the natural landscapes mainly due to regional anthropogenic LCC. In North America, for instance, agriculture was introduced gradually westward with the land demands of new settlers, while in Europe, agriculture is millennia old. The long history of agriculture established the predominantly cultural character of the European landscapes (Klijn 2004). The cultural landscapes range from semi-natural (often-subsidized) grassland in the high Alps and overdrained marshes and wetlands for extensive grazing in Middle and Eastern Europe to completely man-made and high-maintenance polders in The Netherlands.

Worldwide, the extension of cultivated areas changed over time due to changes in local or regional climate, demographic, economic, political and/or natural conditions and agricultural techniques (e.g., introduction of the plough, artificial fertilizer, irrigation). In Europe, for instance, agriculture employed more than 80% of the work force at the beginning of the Middle Age. The population decline due to the pest caused land abandonment (Klijn 2004). Between 1400 and 1800, the agriculturally used area increased. After 1800, the application of agricultural science, and improved education and technology increased the agricultural production. The increase in production and cheap imports from the USA, Canada, Australia and Third World countries led to a decrease of the extent of agriculturally used land. In the first half of the last century, the food demands of the rapidly growing population required to convert formerly poor, less suitable soils into agriculturally used land by means of artificial fertilizer. After the foundation of the European Union (EU), subsidies frequently changed the percentage fraction of what crops were grown. Moreover, the yields per hectare increased due to various economic incentives and policies.

In America, for instance, the settling of the contiguous USA required fertile land for agricultural purpose and wood for construction. Since 1600, these demands led to deforestation of 90% of the virgin forests (Fig. 1.2). In the last century, the vast majority of deforestation in the Tropics occurred for economic reasons. In Amazonia, the deforestation

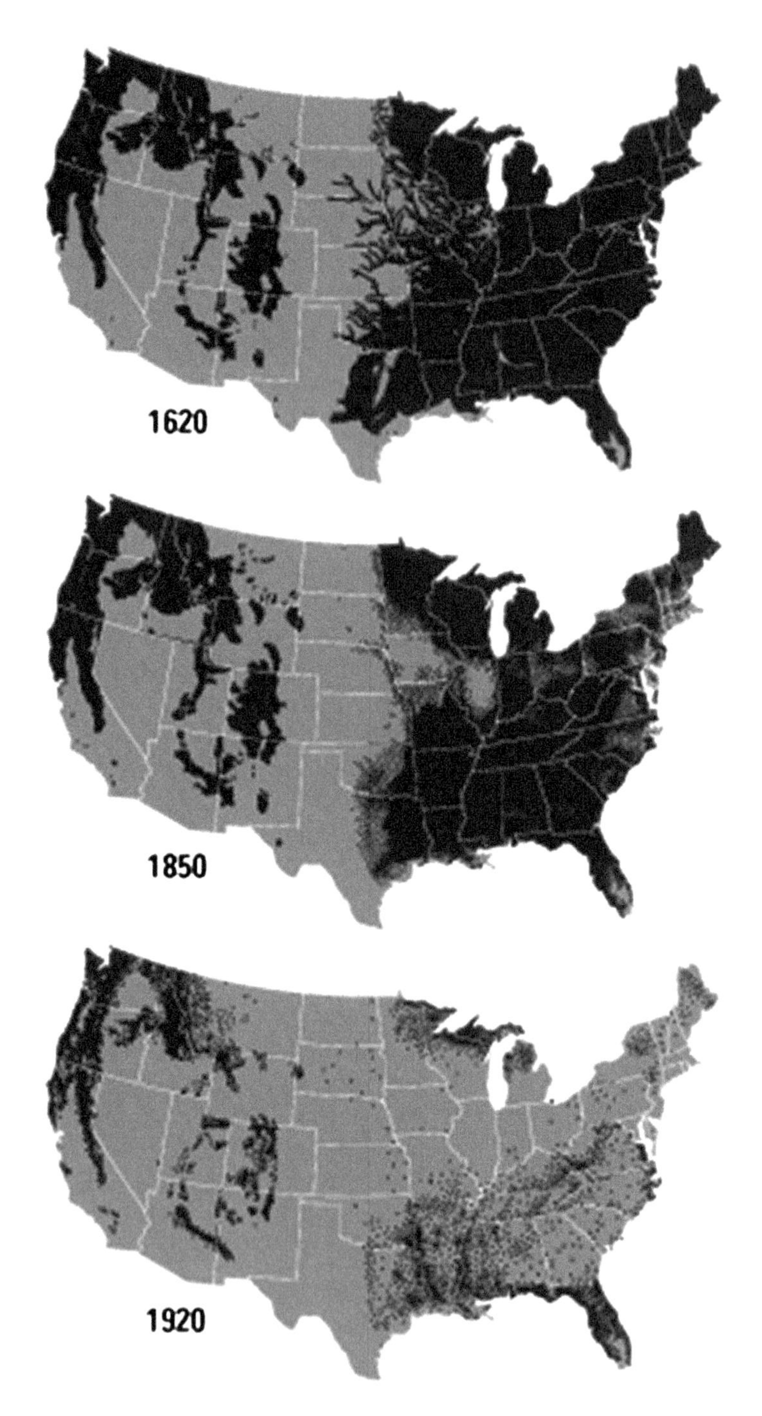

Figure 1.2. Example of historic land-cover changes for the contiguous USA. The *dark gray* area indicates natural forests (Modified after: `http://www.globalchange.umich.edu/globalchange2/current/lectures/deforest/defores9.JPG`)

served to provide land for the poor, while in Asia, powerful families and local governments benefited from harvest concessions. The Tropics of Africa, Latin America, and Asia still have a deforestation rate about 0.8%, 2% and 2% per year, respectively (http://www.globalchange.umich.edu/globalchange2/current/lectures/deforest/deforest.html).

Over time, political changes and subsidence policies led to LCC and/or land-use change. The collapse of the Eastern Block, for instance, meant a shift from large monocultures to individually managed fields. This change in agricultural production increased the heterogeneity of the landscape and altered the pattern of biogenic emissions. Another example for LCC due to political changes is related to the unification of the German states. The new economic conditions led to construction of commercial areas and housing and rearrangement of agriculturally used land in East Germany. Due to the poor quality of the lignite coal, the open-pit mining activities were reduced appreciably. Some of the open-pit mines were closed, and flooded or re-cultivated. These LCC affected local weather (Mölders 1998), anthropogenic and biogenic emissions and, hence, air quality.

An example for the impact of subsidence policies on land-cover and/or land-use is biofuel production. The EU, for instance, set targets of replacing 2% of the union's fossil fuel needs with biofuel. This policy increased the fraction of land-cover used for rapeseed, corn and beet fields. In the USA, grants, loans and tax incentives for biofuel production increased rapidly the fraction of land used for corn production that serves to produce ethanol.

Anthropogenic land-use changes may also occur in response to short-term climate variability. Three to five years of favorable weather for a certain crop (e.g., winter wheat) typically encourage more farmers to sew this crop in the next season. A few years of weather-conditions leading to low yields of a certain crop reduce the fraction of land used to grow this crop. The resulting shortage may cause an increase in the fraction of land used for this crop later on.

1.3 Land-Cover Changes and Weather and Climate

Research on LCC began in the 1960s and 1970s, as the public raised concerns about urban climate, the climate impact of tropical deforestation, and the causes of the great droughts in the Sahel and how to overcome them (Budyko et al. 1971; Potter et al. 1975; Loose and Bornstein

1977). Deforestation was also of concern since a long time, because it reduces the carbon storage and modifies the carbon cycle. Research related to the great droughts in the Sahel led to the hypothesis of biometeorological feedback cycles. The increased albedo due to vegetation loss in the Sahel would trigger sinking motion, additional drying and would foster the arid conditions (Charney et al. 1975).

Today, it is well accepted that LCC can modify local climate conditions significantly (in a statistical sense). However, the full significance of LCC impacts on large-scale (i.e., macro-scale in the classification by Orlanski (1975); Fig. 1.1) climate is still under scrutiny. Land conversion, land-cover and/or land-use changes at all scales have been found to feed back to weather and climate. The impacts of LCC occur highly regionalized and mainly coincide with the population distribution. The anomalies caused by LCC vary with climate region, season, extension of area changed, type of LCC, distance from the LCC and duration over which the LCC persist.

The urban heat-island effect is the most prominent example for impacts of local LCC on local weather and climate. Studies focusing on urban climate showed that urban areas affect the energy budgets (e.g., Kerschgens and Drauschke 1986) and, hence, the distribution of temperature, clouds and precipitation (e.g., Changnon and Huff 1986) and may affect frontal systems (Loose and Bornstein 1977). Examples of regional impacts of LCC are the increased humidity, moderate diurnal temperature cycle, and altered distribution and amount of precipitation after the establishing of large water reservoirs (e.g., Stivari et al. 2005).

Any LCC affect the atmosphere via altered surface characteristics and altered biogenic emissions. The change in surface characteristics modifies the physical conditions of the atmosphere via shifts/changes in biogeophysical processes (e.g., albedo feedback, cloud-evapotranspiration (i.e., the sum of evaporation and transpiration) feedback. The altered biogenic emissions modify the atmospheric composition. If the emitted species are greenhouse gases (GHG), the LCC may affect climate also via altered absorption. Therefore, the increased methane emissions associated with the expansion of rice production are of great public concern.

In case of LCC due to wildfire or slash-and-burn land clearance, the direct process of the LCC affects the atmospheric composition notably at the time the LCC actually occur. Biomass burning, for instance, contributes probably to about 40% of the carbon dioxide, 32% of the carbon monoxide, 20% of the particulates, and 50% of the highly carcinogenic poly-aromatic hydrocarbons emissions worldwide (Sivakumar 2007).

From a scientific point of view, questions related to LCC impacts on the atmosphere are:

- How do LCC affect the atmosphere and its composition and by which mechanisms?
- Do the same LCC produce the same atmospheric responses independent of where they occur?
- What is the critical extent of LCC to produce a significant atmospheric response?
- Do LCC impacts remain locally, or do they affect areas far remote from the LCC? If so, what are the mechanisms causing local and/or remote changes?
- Do simultaneously occurring LCC enhance or diminish each others' responses?
- Do LCC have the same atmospheric responses under altered climate conditions?
- How do the concurrent LCC and GHG impacts on the atmosphere interact with each other?
- Are there possibilities to offset LCC impacts on the atmosphere?
- How does a warming climate affect or trigger natural LCC?
- What are the driving atmospheric forces for natural and anthropogenic LCC?

This book summarizes structures and presents the current knowledge on LCC and land-use change impacts on weather, climate and atmospheric composition, to address these questions. Chapter 2 reviews briefly the theoretical background of atmospheric processes that react sensitive to LCC and the atmospheric anomalies caused by LCC. Chapter 3 presents land-cover impacts on the atmosphere found for current climate conditions from major field campaigns and modeling studies reaching from the local to global scale. Chapter 4 reviews studies on LCC under future climate conditions, including biogeochemical feedbacks; elucidates the challenges related to future LCC and their impact on weather, climate and air quality; elaborates the uncertainties and discusses potential solutions. Chapter 5 presents conclusions and suggests future investigations to explore variables, key parameters and processes that were found to be important. It also poses critical hypotheses that need to be tested in the future.

References

Budyko MI, Drozdov OA, Yudin MI (1971) The impact of economic activity on climate. Sov Geogr Rev Trans 12:666–679

Changnon SA, Huff FA (1986) The urban-related nocturnal rainfall anomaly at St. Louis. J Clim Appl Meteorol 25:1985–1995

Charney J, Stone PH, Quirk WJ (1975) Drought in the Sahara: a biogeophysical feedback mechanism. Science 187:434–435

Kerschgens MJ, Drauschke RL (1986) On the energy budget of a wintry mid-latitude city atmosphere. Contrib Atmos Phys 59:115–125

Klijn JA (2004) Driving forces behind landscape transformation in Europe, from a conceptual approach to policy options. In: Jongman RHD (ed) The New Dimensions of European Landscapes, Springer, Dodrecht, The Netherlands, 201–218

Loose T, Bornstein RD (1977) Observations of mesoscale effects on frontal movement through an urban area. Mon Wea Rev 105:563–571

Mölders N (1998) Landscape changes over a region in East Germany and their impact upon the processes of its atmospheric water-cycle. Meteorol Atmos Phys 68:79–98

Mölders N, Kramm G (2007) Influence of wildfire induced land-cover changes on clouds and precipitation in Interior Alaska – a case study. Atmos Res 84:142–168

Mueller DR, Vincent WF, Jeffries MO (2003) Break-up of the largest Arctic ice shelf and associated loss of an epishelf lake. Geophys Res Lett 30:2031. doi:10.1029/2003gl017931

Munro ND (2003) Small game, the younger dryas, and the transition to agriculture in the southern levant. Mitteilungen der Gesellschaft für Urgeschichte 12:47–64

Orlanski I (1975) A rational subdivision of scales for atmospheric processes. Bull Amer Meteorol Soc 56:527–530

Potter GL, Ellsaesser HW, MacCracken MC, Luther FM (1975) Possible climatic impact of tropical deforestation. Nature 258:697–698

Racine CH, Johnson LA, Viereck LA (1987) Patterns of vegetation recovery after tundra fires in Northwestern Alaska, U.S.A. Arct Alp Res 19:461–469

Rott H, Skvarca P, Nagler T (1996) Rapid collapse of northern Larsen Ice Shelf, Antarctica. Science 271:788–792

Rott H, Rack W, Skvarca P, De Angelis H (2002) Northern Larsen Ice Shelf, Antarctica: further retreat after collapse. Ann Glaciol 34:277–282

Sivakumar MVK (2007) Interactions between climate and desertification. Agric For Meteorol 142:143–155

Stivari S, Oliveira A, Soares J (2005) On the climate impact of the local circulation in the Itaipu Lake area. Clim Change 72:103–121

Vaughan DG, Doake CSM (1996) Recent atmospheric warming and retreat of ice shelves on the Antarctic Peninsula. Nature 379:328–331

Chapter 2

Physical and Chemical Principles

2.1 Energy Balance

The land surface affects the atmosphere via the exchange of heat, matter (moisture, gases, aerosols), and momentum. The exchange of heat and moisture, for instance, are described by the energy and water budgets that are coupled via the water-vapor flux due to evapotranspiration (i.e., the sum of transpiration and evaporation). Coupled energy- and water-budget equations are required to determine the temperature and moisture conditions at the atmosphere-surface interface. In case of low vegetation, for instance, the coupled water- and energy-budget equations for the vegetation-soil-atmosphere interface read

$$F_1(\mathbf{S}) = R^{\downarrow}_{\mathrm{sf}} - R^{\uparrow}_{\mathrm{sf}} + R^{\downarrow}_{\mathrm{lf}} - R^{\uparrow}_{\mathrm{lf}} - R^{\downarrow}_{\mathrm{sg}} + R^{\uparrow}_{\mathrm{sg}} - R^{\downarrow}_{\mathrm{lg}} + R^{\uparrow}_{\mathrm{lg}} - H_{\mathrm{f}} - L_{\mathrm{v}}E_{\mathrm{f}} = 0 \quad (2.1)$$

$$F_2(\mathbf{S}) = R^{\downarrow}_{\mathrm{sg}} - R^{\uparrow}_{\mathrm{sg}} + R^{\downarrow}_{\mathrm{lg}} - R^{\uparrow}_{\mathrm{lg}} - H_{\mathrm{g}} - L_{\mathrm{v}}E_{\mathrm{g}} + G = 0 \quad (2.2)$$

$$F_3(\mathbf{S}) = I_{\mathrm{g}} + W_{\mathrm{soil}} - E_{\mathrm{g}} = 0 \quad (2.3)$$

where the subscripts f and g denote to the surfaces of foliage and soil. Here, $R^{\downarrow}_{\mathrm{sf}}$, $R^{\uparrow}_{\mathrm{sf}}$, $R^{\downarrow}_{\mathrm{lf}}$, $R^{\uparrow}_{\mathrm{lf}}$, $R^{\downarrow}_{\mathrm{sg}}$, $R^{\uparrow}_{\mathrm{sg}}$, $R^{\downarrow}_{\mathrm{lg}}$, and $R^{\uparrow}_{\mathrm{lg}}$ are the downward (↓) and upward (↑) directed fluxes of shortwave (index s) and long-wave (index l) radiation. Furthermore, H, $L_{\mathrm{v}}E$, G, W_{soil}, I_{g}, and L_{v} are the fluxes of sensible and latent heat, the soil heat and water fluxes at the surface, infiltration and the latent heat of vaporization, respectively.

N. Mölders, *Land-Use and Land-Cover Changes*, Atmospheric and Oceanographic Sciences Library 44, DOI 10.1007/978-94-007-1527-1_2,

For other or more surfaces, similar sets of coupled equations can be formulated. In the above notation, F_1, F_2, and F_3 are the "vectors" of the budget equations, and $\mathbf{S}$ is the vector of the surface values.

Such coupled nonlinear budget equations can be solved simultaneously by a Newton–Raphson iteration procedure of the first-order. In case of Eqs. 2.1–2.3, this procedure reads (e.g., Kramm et al. 1996)

$$\mathbf{S}^{(n+1)} = \mathbf{S}^{(n)} - \mathrm{DF}\left(\mathbf{S}^{(n)}\right)^{-1} F\left(\mathbf{S}^{(n)}\right) \tag{2.4}$$

where $\mathbf{S} = [T_\mathrm{f}\ T_\mathrm{g}\ \eta_\mathrm{g}]^\mathrm{T}$, $F(\mathbf{S}) = [F_1(\mathbf{S})\, F_1(\mathbf{S})\, F_3(\mathbf{S})]^\mathrm{T}$, and $\mathrm{DF}(\mathbf{S})$ and $\mathrm{DF}(\mathbf{S})^{-1}$ are the functional matrix, and its inverse, respectively, and the superscript T denotes the transpose. Furthermore, T_f, T_g, and η_g are the temperature at the foliage and ground and the soil volumetric water content at the soil surface. Typically, the iteration procedure will stop if a sufficient accuracy of $|F_1(\mathbf{S}^k)| < 0.1\,\mathrm{W\,m^{-2}}$, $|F_2(\mathbf{S}^k)| < 0.1\,\mathrm{W\ m^{-2}}$, and $|F_3(\mathbf{S}^k)| < 10^{-7}$ is achieved after the kth iteration. Usually, these criteria are fulfilled after five to seven iterations. To save computational time, these types of coupled equations are often decoupled by assuming that on short time scales, the moisture and heat fluxes negligibly change temperature and moisture, respectively. The soil water-vapor flux can also be used to estimate the bulk resistance of a soil layer against transfer of trace gases (e.g., methane emission from rice fields or the active layer of permafrost soils).

Analogous to the energy and water budget of the soil–vegetation system, energy and water budgets can be formulated for a system consisting of snow (second subscript s) and underlying soil (subscript g)

$$\begin{aligned} F_1(\mathbf{S}) = {} & R_\mathrm{ss}^{\downarrow} - R_\mathrm{ss}^{\uparrow} + R_\mathrm{ls}^{\downarrow} - R_\mathrm{ls}^{\uparrow} - H_\mathrm{s} - L_\mathrm{s}E_\mathrm{s} + G_\mathrm{snow} \\ & + P_\mathrm{H} - R_\mathrm{sg}^{\downarrow} + R_\mathrm{sg}^{\uparrow} = 0 \end{aligned} \tag{2.5}$$

$$F_2(\mathbf{S}) = R_\mathrm{sg}^{\downarrow} - R_\mathrm{sg}^{\uparrow} + G - G_\mathrm{snow} = 0 \tag{2.6}$$

$$F_3(\mathbf{S}) = P + S - E_\mathrm{s} - W_\mathrm{soil} = 0 \tag{2.7}$$

where $\mathbf{S} = [T_\mathrm{snow,surf}\ T_\mathrm{g}\ \eta_\mathrm{g}]^\mathrm{T}$ is the vector of the surface values and $T_\mathrm{snow,surf}$ is the temperature of the snow surface. Here, $R_\mathrm{ss}^{\downarrow}$, $R_\mathrm{ls}^{\downarrow}$, $R_\mathrm{ss}^{\uparrow} = \alpha_\mathrm{snow}R_\mathrm{ss}^{\downarrow}$, and $R_\mathrm{ls}^{\uparrow}$ are the downward and upward directed fluxes of shortwave and long-wave radiation. Furthermore, P and S are the liquid- and solid- phase precipitation, and L_s is the latent heat of sublimation. Since snow is semitransparent to shortwave radiation, the downward and upward directed fluxes of shortwave radiation through the snowpack of thickness z_snow to and from the ground read

$$R_{\mathrm{sg}}^{\downarrow} = R_{\mathrm{ss}}^{\downarrow} \exp\left(-k_{\mathrm{ext}} z_{\mathrm{snow}}\right) \tag{2.8}$$

$$R_{\mathrm{sg}}^{\uparrow} = \alpha_{\mathrm{g}} R_{\mathrm{sg}}^{\downarrow} \tag{2.9}$$

where k_{ext} is the extinction coefficient of snow. Non-supercooled rain provides heat, P_{H}, to the snowpack. The heat flux into the snowpack can be determined as

$$G_{\mathrm{snow}} = -\lambda_{\mathrm{snow}} \frac{T_{\mathrm{snow,surf}} - T_{\mathrm{g}}}{z_{\mathrm{snow}}} - L_{\mathrm{v}} \rho_{\mathrm{w}} k_{\mathrm{v}} \frac{(q_{\mathrm{snow,surf}} - q_{\mathrm{g}})}{z_{\mathrm{snow}}} \tag{2.10}$$

where λ_{snow} and k_{v} are the thermal conductivity of snow and molecular diffusion coefficient of water vapor within air-filled pores of the snowpack.

Snow emissivity, $\varepsilon_{\mathrm{snow}}$, depends on snow age. Snow albedo, α_{snow}, depends on precipitation history, snow depth, radiation, sun angle, wavelength, grain size and type, liquid water content of the snowpack, meteorological conditions, and air-pollution effects (Wiscombe and Warren 1980). Soil albedo, α_{g}, depends on soil type and moisture (Pielke 2002).

2.1.1 Latent and Sensible Heat Fluxes

The latent heat fluxes are related to evapotranspiration via the latent heat released/consumed during the phase-transition process. Evapotranspiration depends on meteorological conditions, soil-physical and geologic conditions and characteristics, and biological characteristics. Meteorological conditions affecting evapotranspiration are wind-speed, humidity, air temperature, and photosynthetic active radiation (PAR). Soil-physical and geologic conditions and characteristics important for evaporation are soil moisture and temperature, soil type, aeration, fertilizer, and biological and soil-chemical processes. Important biological characteristics are, among others, vegetation height, health of vegetation, amount of roots at soil levels with sufficient plant-available water, competition or interaction with roots of other species, and stomatal resistance. The water extraction by roots and the following transpiration act as a sink for soil water and have to be considered in the calculation of soil-moisture states. The sensible heat fluxes depend on meteorological conditions and surface temperatures.

To consider the different fluxes at the soil–vegetation–atmosphere interface, the effects of bare and vegetation-covered soil on evapotranspiration are weighted by the shielding factor σ_{f} ($0 \leqslant \sigma_{\mathrm{f}} \leqslant 1$). The shielding factor describes the degree to which foliage prevents shortwave radiation from reaching the ground (Fig. 2.1). The governing flux equations for water vapor, E, and sensible heat, H, at the surfaces of foliage and soil are given by

$$E_{\mathrm{f}} = -\sigma_{\mathrm{f}}\rho_{\mathrm{a}}\left\{\frac{1}{r_{\mathrm{mt,f}}}(q_{\delta}-q_{\mathrm{f}})-\frac{1}{r_{\mathrm{mt,fg}}}(q_{\mathrm{f}}-q_{\mathrm{g}})\right\} \tag{2.11}$$

$$H_{\mathrm{f}} = -\sigma_{\mathrm{f}}c_{\mathrm{p}}\rho_{\mathrm{a}}\left\{\frac{1}{r_{\mathrm{mt,f}}}(\Theta_{\delta}-T_{\mathrm{f}})-\frac{1}{r_{\mathrm{mt,fg}}}(T_{\mathrm{f}}-T_{\mathrm{g}})\right\} \tag{2.12}$$

$$E_{\mathrm{g}} = -\rho_{\mathrm{a}}\left\{\frac{1-\sigma_{\mathrm{f}}}{r_{\mathrm{mt,g}}}(q_{\delta}-q_{\mathrm{g}})+\frac{\sigma_{\mathrm{f}}}{r_{\mathrm{mt,fg}}}(q_{\mathrm{f}}-q_{\mathrm{g}})\right\} \tag{2.13}$$

$$H_{\mathrm{g}} = -c_{\mathrm{p}}\rho_{\mathrm{a}}\left\{\frac{1-\sigma_{\mathrm{f}}}{r_{\mathrm{mt,g}}}(\Theta_{\delta}-T_{\mathrm{g}})+\frac{\sigma_{\mathrm{f}}}{r_{\mathrm{mt,fg}}}(T_{\mathrm{f}}-T_{\mathrm{g}})\right\} \tag{2.14}$$

The resistances of the molecular turbulent layer close to the foliage and soil and the molecular turbulent region between the foliage and soil surface (index fg) against the transfer of heat and matter are $r_{\mathrm{mt,f}}$, $r_{\mathrm{mt,g}}$, and $r_{\mathrm{mt,fg}}$, respectively. Furthermore, c_{p} and ρ_{a} are the specific heat at constant pressure and density of air. According to Ohm's law of electrostatics, the eddy fluxes of water vapor and sensible heat across the turbulent region of the atmospheric surface layer (ASL) between δ and the reference height z_{R} (level of measurements or model layer) read

$$E_{\mathrm{t}} = -\frac{\rho_{\mathrm{a}}}{r_{\mathrm{t}}}(q_{\mathrm{R}}-q_{\delta}) \tag{2.15}$$

$$H_{\mathrm{t}} = -\frac{\rho_{\mathrm{a}}}{r_{\mathrm{t}}}(c_{\mathrm{p}}(\Theta_{\mathrm{R}}-\Theta_{\delta})) \tag{2.16}$$

where q_{R} and Θ_{R} are the specific humidity and the potential temperature at z_{R}, respectively. The quantities

$$\Theta_{\delta} = \frac{r_{\mathrm{mt,H}}r_{\mathrm{t}}}{r_{\mathrm{mt,H}}+r_{\mathrm{t}}}\left\{\frac{1-\sigma_{\mathrm{f}}}{r_{\mathrm{mt,g}}}T_{\mathrm{g}}+\frac{\sigma_{f}}{r_{\mathrm{mt,f}}}T_{\mathrm{f}}+\frac{\Theta_{\mathrm{R}}}{r_{\mathrm{t}}}\right\} \tag{2.17}$$

and

$$\begin{aligned} q_{\delta} = \frac{r_{\mathrm{mt,E}}r_{\mathrm{t}}}{r_{\mathrm{mt,E}}+r_{\mathrm{t}}}&\left\{\left\{\frac{1-\sigma_{\mathrm{f}}}{r_{\mathrm{mt,g}}}+\frac{\frac{\sigma_{\mathrm{f}}}{r_{\mathrm{mt,fg}}}}{\left(\frac{1}{r_{\mathrm{st}}}+\frac{1}{r_{\mathrm{mt,fg}}}+\frac{1}{r_{\mathrm{mt,f}}}\right)r_{\mathrm{mt,f}}}\right\}q_{\mathrm{g}}\right. \\ &\left.+\frac{\frac{\sigma_{\mathrm{f}}}{r_{\mathrm{st}}}}{\left(\frac{1}{r_{\mathrm{st}}}+\frac{1}{r_{\mathrm{mt,fg}}}+\frac{1}{r_{\mathrm{mt,f}}}\right)r_{\mathrm{mt,f}}}q_{\mathrm{i}}+\frac{q_{\mathrm{R}}}{r_{\mathrm{t}}}\right\} \end{aligned} \tag{2.18}$$

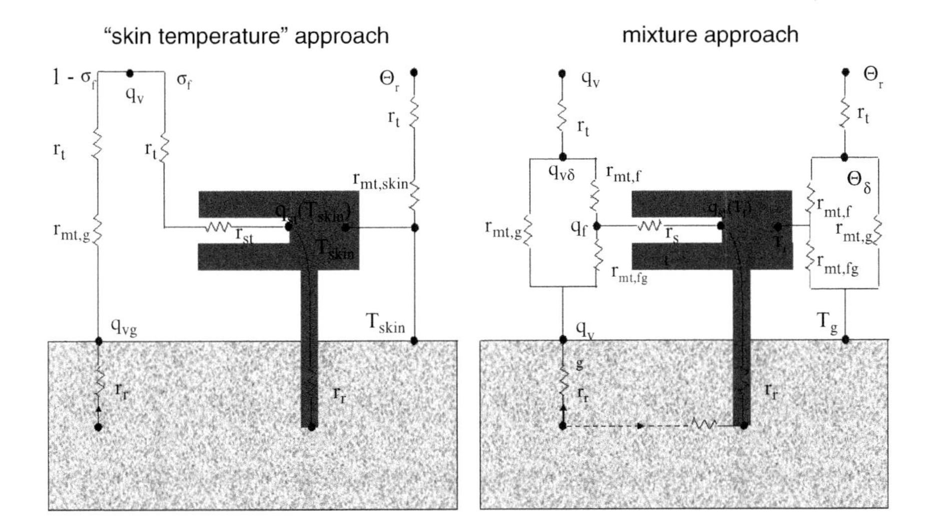

Figure 2.1. Schematic view of the big-leaf big-stomata resistance network approach once assuming a skin temperature for the entire grid-cell and once a mixture approach that allows different temperatures for the foliage and the ground. Here r_{r}, r_{t}, r_{st}, r_{mt}, $r_{\mathrm{mt,skin}}$, T_{skin}, q_{st}, θ_r, q_{v}, are the resistance for water-uptake by roots, turbulent resistance, stomatal resistance, molecular turbulent resistance, molecular turbulent resistance for the skin surface, skin temperature, specific humidity in the stomata, potential temperature and specific humidity at reference height, respectively. For further symbol explanation, see text

are the potential temperature and specific humidity at height δ close above the foliage. Furthermore, T_{f},

$$q_{\mathrm{f}} = \frac{1}{\frac{1}{r_{\mathrm{st}}} + \frac{1}{r_{\mathrm{mt,fg}}} + \frac{1}{r_{\mathrm{mt,f}}}} \left\{ \frac{q_{\mathrm{i}}}{r_{\mathrm{st}}} + \frac{q_{\mathrm{g}}}{r_{\mathrm{mt,fg}}} + \frac{q_{\delta}}{r_{\mathrm{mt,f}}} \right\}, \tag{2.19}$$

T_{g}, and q_{g} are the surface temperature and specific humidity of foliage and the soil at the ground surface, respectively. Typically, the specific humidity within the stomata cavities $q_{\mathrm{i}} = q_{\mathrm{i}}(T_{\mathrm{f}})$ is assumed to be at saturation. Eqs. 2.17–2.19 can be derived by applying Kirchhoff's law of electrostatics (e.g., Fig. 2.1).

The resistance of that turbulent region against the transfer of heat and matter reads

$$r_{\mathrm{t}} = \frac{1}{u_{*}\kappa} \left(\ln \frac{z_{\mathrm{R}} - d_0}{\delta - d_0} - \Psi_{\mathrm{h}} \left(\zeta_{\mathrm{R}}, \zeta_{\delta} \right) \right) \tag{2.20}$$

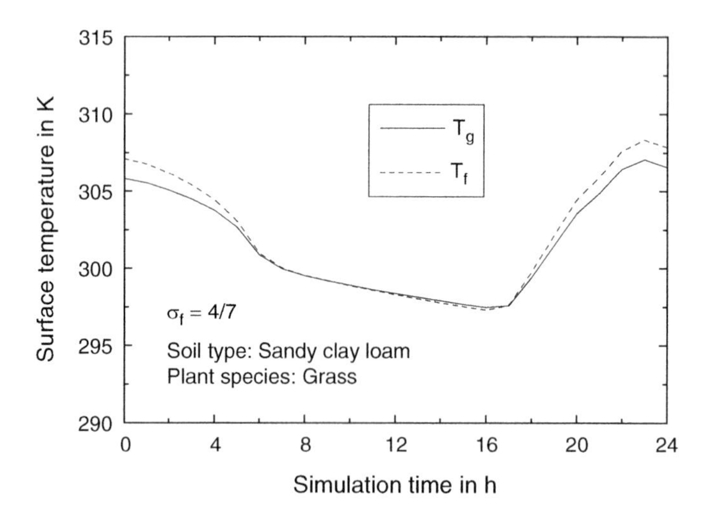

Figure 2.2. Example of differences in soil surface and foliage-surface temperature as obtained for grass covering sandy clay loam by about 57%. The simulation started at noon (From Mölders 1999)

where u_*, κ, and $\zeta = (z - d_0)/L$ are the friction velocity, von Kármán constant, and Obukhov number, respectively. Furthermore, L and d_0 are the Obukhov-stability length and zero-plane displacement, and z stands for z_R or δ. The turbulent resistance, r_t, depends on the thermal stratification of the turbulent part of the ASL. Typically, r_t is described by non-dimensional integral stability functions for heat and matter. The quantities $r_\mathrm{mt,E}$ and $r_\mathrm{mt,H}$ are defined by

$$r_\mathrm{mt,E} = \left(\frac{1-\sigma_\mathrm{f}}{r_\mathrm{mt,g}} + \frac{\left(\frac{1}{r_\mathrm{st}} + \frac{1}{r_\mathrm{mt,fg}}\right)\sigma_\mathrm{f}}{\left(\frac{1}{r_\mathrm{st}} + \frac{1}{r_\mathrm{mt,fg}} + \frac{1}{r_\mathrm{mt,f}}\right) r_\mathrm{mt,f}} \right)^{-1} \tag{2.21}$$

$$r_\mathrm{mt,H} = \left(\frac{1-\sigma_\mathrm{f}}{r_\mathrm{mt,g}} + \frac{\sigma_\mathrm{f}}{r_\mathrm{mt,f}} \right)^{-1} \tag{2.22}$$

Many models do not distinguish between the surface temperature of the ground and foliage (and/or snow) and use a common skin temperature instead to reduce the complexity of the above equations. However, doing so may introduce errors, as the temperatures of the different surfaces may differ by several K (e.g., Fig. 2.2). The magnitude of error depends on the difference in thermal and hydrological properties among the surface types considered and varies over the diurnal course.

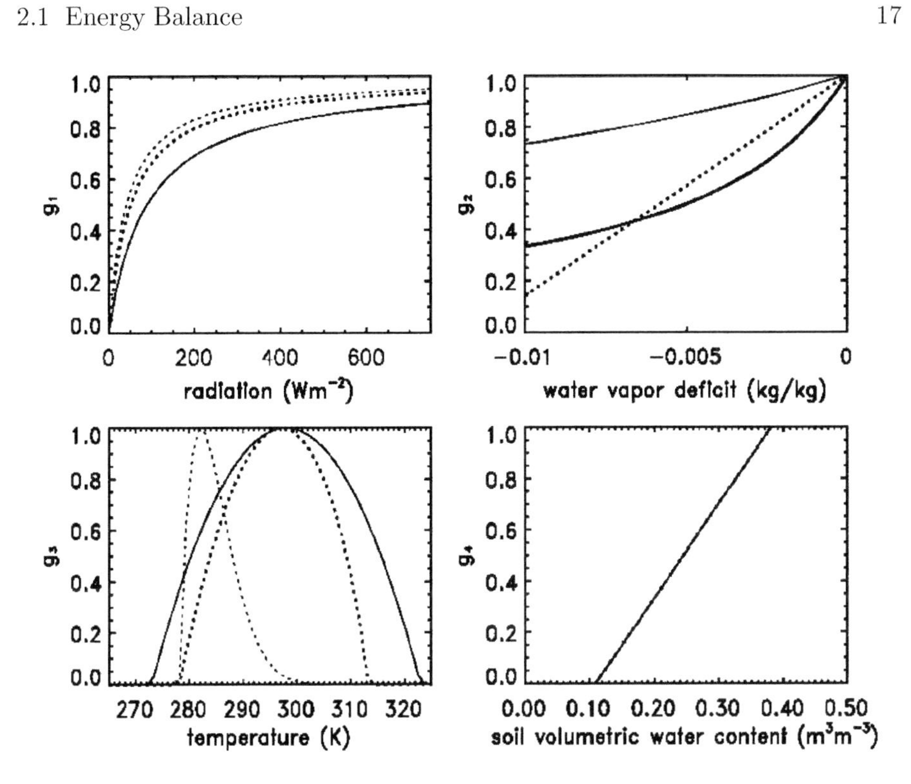

Figure 2.3. Comparison of Jarvis-type correction functions for correcting for the effects of photosynthetic active radiation, water-vapor deficit, near-surface air temperature, and soil volumetric water content as used in the hydro-thermodynamic soil vegetation scheme (HTSVS; Kramm et al. 1996; Mölders et al. 2003) and the Orgeon State University land surface model (OSULSM; Chen and Dudhia 2000)

Most models assume bulk-stomatal resistance to describe transpiration. Frequently, Jarvis (1976) type correction factors consider the sensitivity of the bulk-stomatal resistance

$$r_{\mathrm{st}} = \frac{r_{\mathrm{st,min}}}{g_{\mathrm{R}}\,(\mathrm{PAR})\, g_{\mathrm{q}}\,(\Delta q)\, g_{\mathrm{T}}\,(T_{\mathrm{f}})\, g_{\eta}\,(\Delta \eta)\, g_{\mathrm{CO_2}}\,(\mathrm{CO_2})} \tag{2.23}$$

to PAR, specific humidity deficit between leaf and ambient air $\Delta q = q(T_{\mathrm{f}}) - q_{\delta}$, foliage temperature T_{f}, soil moisture deficit at various soil levels $\Delta\eta$, and the volumetric carbon dioxide (CO_2) concentration. The correction functions g range from 0 (no impact) to 1 (full impact). Here, $r_{\mathrm{st,min}}$ is the plant-specific minimum stomatal resistance.

Such semiempirical correction functions have been derived from observations. Thus, correction functions vary among authors (e.g., Fig. 2.3). Differences in calculated correction factors at same ambient conditions reflect the regions, soil, and vegetation types for which the correction functions were determined and the assumptions made in determining them. The main differences with respect to temperature-correction functions, for instance, are in the assumed minimum and maximum temperatures marking the temperature range for which stomata should always be open, and the temperature, at which r_{st} becomes minimal. The main differences in correction functions for soil moisture result from whether they account for the soil-water deficit in the root zone or not. Correction functions that consider soil-water uptake by roots may differ due to the assumptions on the root distribution, maximum root length, the share of potential supply of available water from different layers, and parameters for the internal vascular resistance per unit length of roots and wilting point.

The correction functions for the sensitivity to CO_2 vary with vegetation type. Hints exist that plants may respond to increased atmospheric CO_2 concentration by a long-term increase of the stomatal resistance (Dingman 1994). See Rosenberg et al. (1990) for a more detailed review on correction functions and their impacts on transpiration calculated therewith.

The fluxes of water vapor and sensible heat over snow can be determines by

$$E_{\mathrm{s}} = -\frac{\rho_{\mathrm{a}}}{r_{\mathrm{mt,snow}} + r_{\mathrm{t}}}(q_{\mathrm{R}} - q_{\mathrm{snow,surf}}) \tag{2.24}$$

$$H_{\mathrm{s}} = -\frac{c_{\mathrm{p}}\rho_{\mathrm{a}}}{r_{\mathrm{mt,snow}} + r_{\mathrm{t}}}(\Theta_{\mathrm{R}} - T_{\mathrm{snow,surf}}) \tag{2.25}$$

where $T_{\mathrm{snow,surf}}$, $q_{\mathrm{snow,surf}}$, and $r_{\mathrm{mt,snow}}$ are the temperature, specific humidity, and resistance against molecular turbulent fluxes at the snow surface. Typically, $q_{\mathrm{snow,surf}}$ is set equal to the specific humidity at saturation with respect to ice.

2.1.2 Surface Heterogeneity

The exchange of water, energy, trace gases, and momentum at the surface-atmosphere interface differs among surface types. In models, several land-cover/use types may occur within the area represented by a grid-cell. Consequently, these various land-cover/use types are of subgrid-scale with respect to the resolution of the model. Typically, the surface within a grid-cell has heterogeneity at various scales. On the microscale, for instance, heterogeneity occurs between the vegetation and the ground or between vegetation sticking out of snow and the snow. This heterogeneity within a patch of seemingly same type (e.g., grass) is called inhomogeneity. Inhomogeneity can be considered by a mixture approach (Deardorff 1978) that determines tightly coupled energy balances for the different surface types (e.g., vegetation and soil) like Eqs. 2.1–2.25. Theoretically, we can extend these equations for more than two different surface types. Doing so, however, increases the number of coupled differential equations for which solving the equation system iteratively becomes computationally challenging.

Therefore, scientists developed various methods for describing the exchange of momentum, heat, energy, and trace gases for heterogeneous surfaces. These methods encompass averaging of surface properties, statistical–dynamical approaches, and various types of mosaic approaches. They use some forms of Eqs. 2.1–2.10 to calculate the surface fluxes. For further details on the strategies to consider subgrid-scale heterogeneity, see Mahrt and Sun (1995), Mölders and Raabe (1996), as well as Giorgi and Avissar (1997).

The simplest approach to calculate the exchange of fluxes for a heterogeneous surface-atmosphere interface is the use of averaged – so-called effective – surface parameters (e.g., Dolman 1992; Lhomme 1992). This means that in the above equations, the area fluxes are calculated using averaged surface parameters rather than surface type–dependent parameters. Theoretical studies, however, imply that averaging surface parameters often fails to provide accurate area averages of the fluxes (Giorgi and Avissar 1997).

Another computationally simple approach is the strategy of dominant land-cover/use. It assumes that the dominating land-cover/use type within a grid-cell is representative for the entire grid-cell. This strategy ignores all heterogeneity within the grid-cell (Fig. 2.4). Resulting errors are greatest for large numbers of different land-cover/use within the grid-cell and for strong differences between the surface properties of the dominant land-cover/use and the neglected other land-cover/use types within the grid-cell.

On the mesoscale, in addition to inhomogeneity, heterogeneity exists between adjacent areas of different surface conditions (e.g., snow-covered vs. snow-free areas). This kind of heterogeneity is often treated by some kind of mosaic approach. Herein, each grid-cell area is divided into n homogeneous subregions – called patches – of different surface types (Fig. 2.4). Most mosaic approaches assume that horizontal fluxes between different patches are small, compared to the corresponding vertical fluxes, for which patches of same type can be arranged into a single

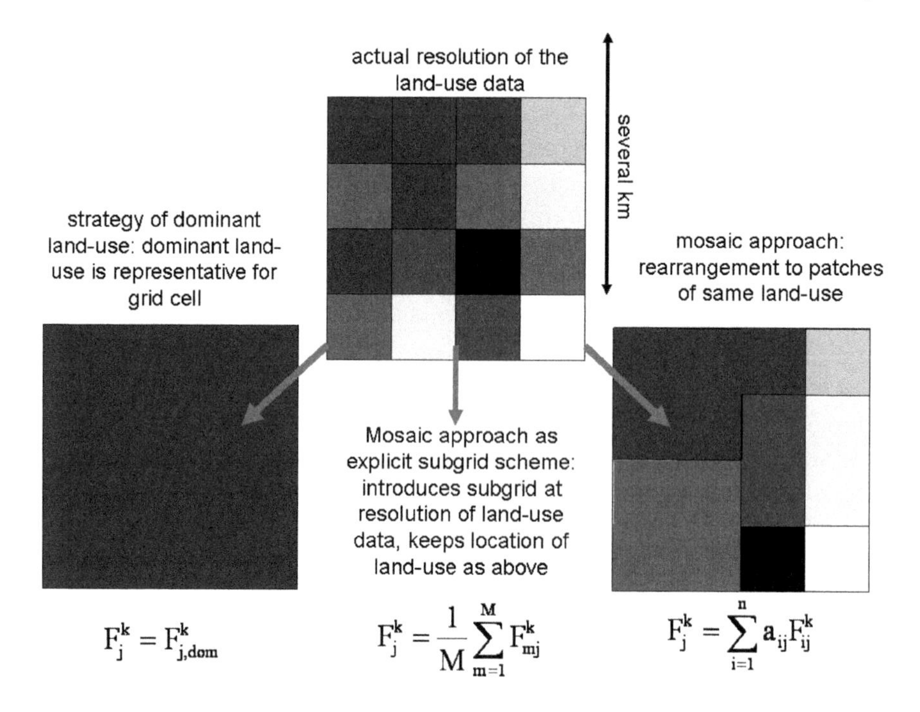

Figure 2.4. Schematic view of different strategies to treat land-surface heterogeneity within a model grid-cell. The explicit subgrid strategy (*middle*) features a mosaic approach wherein the patches keep their location and the micrometeorological and soil conditions, vegetation and soil type are considered at these locations

patch. This assumption neglects advective effects accompanied with internal boundary layers, dynamical effects, or directed flow related to land-surface heterogeneity. Such directed flows can result from topography or mesoscale circulations (e.g., channeling, mountain-valley, circulations, land–sea breezes, snow breeze, vegetation breeze).

In most mosaic approaches, for each of the patches, the energy and water fluxes, soil temperature, and moisture states are determined, ignoring the original location of the individual patches. Mean atmospheric field quantities (in the immediate vicinity of the surface and at reference height) as well as the individual soil moisture and temperatures serve to calculate the energy, trace gas, and water fluxes. The feedback from the surface to the atmosphere over the total area of the grid-cell is accomplished by determining area-weighted fluxes that describe the exchange at the surface–atmosphere interface. The computational expenses of a mosaic approach using the patch technique increase with the number of different patches within the grid-cell.

Variations of the mosaic approach are the mixture approach, the explicit sub-grid strategy and the blending height concept. The explicit subgrid strategy maintains the location of the individual patches (Seth et al. 1994). The explicit subgrid strategy is advantageous for strongly varying, complex terrain as it permits to have same land-cover/use in combination with different elevation, soil type and moisture, wind channeling, etc. Its main disadvantages are its huge storage and computational demands.

The blending height concept (e.g., Claussen 1990) estimates effective transfer coefficients by averaging the turbulent scales of velocity and concentrations based on a blending height. The blending height represents a height up to which vertical diffusion can affect the mean flow. It is the height, at which the sum of the mean flow's deviations from the local equilibrium and horizontal homogeneity is minimal. Above this height, a heterogeneous surface fails to modify the flow individually. Instead, overall fluxes and mean profiles represent the surface condition of the area.

Analysis of field experiments showed that assuming spatially constant atmospheric variables and spatially varying surface conditions closely approximate the area composite fluxes (Mahrt and Sun 1995). Mosaic approaches follow this principle.

Fluxes from patchy surfaces differ strongly from fluxes from homogeneous (one surface type) surfaces (e.g., Fig. 2.5). The Bowen ratios B (ratio of sensible to latent heat flux) shift toward lower values when subgrid-scale cool, moist areas are considered (e.g., parks and water meadows in conurbations, oasis, irrigated areas in deserts) for regions dominated by relatively dry, warmer surfaces. The opposite is true for subgrid-scale warm, dry areas (e.g., small settlements in agriculturally used land) in regions dominated by relatively cooler, moister surfaces.

The effective differences in results obtained by the mixture and mosaic approach are small over a wide range of conditions (Koster and Suarez 1992). The results of mosaic approaches can be very sensitive to the assumptions on the degree of heterogeneity. The same vegetation, for instance, must necessarily not occur on the same soil type, or soil states may differ for same soil or vegetation type. Inclusion or neglecting of such aspects of heterogeneity can appreciably affect simulated fluxes and state variables (Mölders et al. 1996).

Modern climate models and earth-system models use some kind of mosaic approach to consider the subgrid-scale heterogeneity of land-cover/use (Bonan 2008). Many mesoscale research models consider subgrid-scale heterogeneity of land-cover/use when they use large horizontal grid spacing (Giorgi and Avissar 1997).

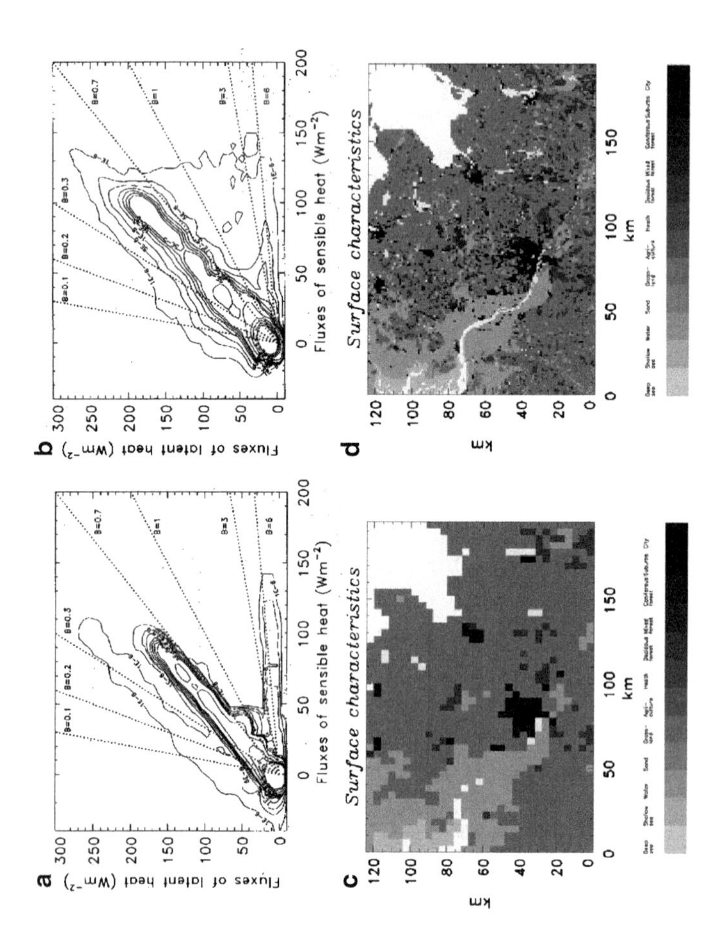

Figure 2.5. Twenty-four-hour frequency of pairs of fluxes of sensible and latent heat as obtained for an area of 128 km in North-South and 200 km in West-East direction around Hamburg, Germany. Simulations were preformed with 4 km horizontal grid spacing. For (**a**), the simulation was run assuming each grid-cell as homogeneously covered by the dominant land-use type as shown in (**c**). For (**b**), the simulation was run using a mosaic approach realized in form of an explicit subgrid strategy with a horizontal subgrid spacing of 1 km as shown in (**d**) (Modified after Mölders 2001)

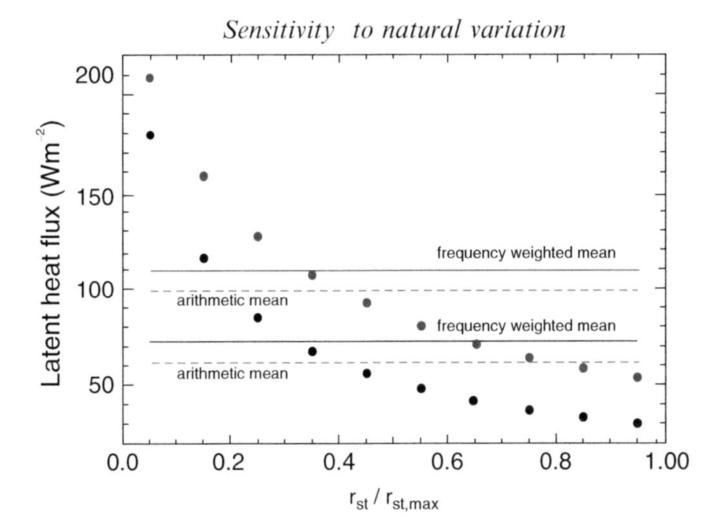

Figure 2.6. Latent heat fluxes vs. relative stomatal resistance, $r_{st}/r_{st,max}$ alternatively assuming maximum stomatal resistances $r_{st,max}$ of 500 s/m (*gray*) and 1,000 s/m (*black*). The solid and dashed lines represent the latent heat flux weighted by the relative stomatal resistance and simulated by use of the arithmetic average of $r_{st}/r_{st,max}$ (From Mölders 2001)

The surface parameters used in the above equations stem from measurements and are not natural constants. Instead, individual measurements differ from each other often by more than an order of magnitude. Using a mean value may cause huge uncertainty in simulated temperature and moisture states as well as fluxes through error propagation (Mölders et al. 2005). Thus, some authors suggested calculating the surface fluxes by including the frequency of the parameter values rather than just using mean values of parameters (e.g., Avissar 1993).

Figure 2.6 exemplary compares the area-aggregated fluxes obtained by use of arithmetic averages of relative stomatal resistance gained from more than 1,000 measurements performed over an area of 3 m radius in a potato field under relative constant atmospheric conditions (Avissar 1993). The latent heat fluxes determined by use of the arithmetic mean of stomatal resistance underestimate the area-mean latent heat fluxes calculated physically correctly from the frequency-weighted fluxes gained for the range of stomatal resistances. The latent heat fluxes determined by use of the arithmetic mean of stomatal resistance and frequency-weighting of fluxes differ about 12% and 18% for a maximum stomatal resistance of 1,000 s/m and 500 s/m, respectively (Mölders 2001).

2.2 Clouds and Precipitation

Evapotranspiration and/or sublimation from land and water surfaces provide water vapor to the atmosphere. If the air becomes supersaturated, cloud droplets and/or ice crystals form by heterogeneous nucleation on aerosols functioning as cloud condensation nuclei (CCN) and/or ice nuclei (IN). These nuclei exist in the atmosphere from natural and/or anthropogenic sources. Natural CCN sources are volcanic eruptions, dust, sea-spray salt, bacteria, and wildfires. Plants (e.g., pollen), biomass burning, and combustion of bio- and fossil fuel provide organic CCN. Gas-to-particle conversion creates CCN in the atmosphere. Organic IN stem from seawater plankton or decayed plant leaves. In snow crystals, about 87% of the IN are clay minerals. Since a variety of the above sources for CCN and IN is land-cover/use related, the distribution of CCN and IN and cloud formation become sensitive to land-cover/use changes.

Reaching supersaturation activates water-soluble CCN. If the surface tension between the CCN surface and the water is small enough, a water film can form around the CCN. For thermodynamic reasons, a critical radius exists above which an initially formed embryo droplet can survive. Any further droplet growth decreases the total Gibbs free energy of the system. This is also true for embryo-ice crystals. Since the saturation-vapor pressure of pure water exceeds that of a solute, the nucleation process is most efficient for soluble CCN. The first embryo droplets form on the greatest and most soluble CCN. Thus, land-surface emissions of soluble aerosols may enhance the likelihood of nucleation.

Once an embryo droplet (ice crystal) formed, it can grow by water-vapor diffusion known as condensation (deposition). Water-vapor diffusion directed away from the droplet (crystal) leads to evaporation (sublimation). Water-vapor diffusion depends on the temperature and humidity of the environment and the radius of the embryo droplet or ice crystal. Since only few IN exist, the size distribution of ice crystals is spread over few large particles. Since large ice particles are the best nucleus for further diffusion growth, the ice particles develop quickly fall-speeds of tens of cm/s.

As the droplets/crystals become great enough, they start to settle. Since the droplets and/or ice crystals differ in size, their fall velocities differ, too. The quicker particles collect the smaller ones. In the collection, particles of different phase may be involved. We speak of coalescence for collection of droplets by droplets, riming for collection of cloud droplets by ice crystals, and aggregation for collection of crystals by crystals or snowflakes by snowflakes. Since the terminal velocity of any particle (i.e., the non-accelerated final velocity) grows with increasing particle radius, changes in particle size modify the efficiency of precipitation formation.

At temperatures between the freezing point and $-35\,^{\circ}$C to $-40\,^{\circ}$C super-cooled water and ice may coexist. Since the binding forces of ice exceed those of water, the saturation-vapor pressure of water exceeds that of ice. Under supersaturated conditions with respect to ice, but subsaturated conditions with respect to water, water-vapor diffusion yields that ice crystals grow at the cost of cloud droplets (Bergeron–Findeisen process). The lower saturation-vapor pressure over ice than water also means that the cold path of precipitation formation via the ice phase is more efficient than the warm path that involves only the liquid phase. Via the cold path of precipitation formation, more water vapor can transform to precipitation than via the warm path.

The saturation pressure over a concave surface exceeds that over a convex surface, for which more complicated dendrite-like ice crystals convert into less complex shapes. This means the degree of supersaturation determines the ice-crystal type with consequences for terminal velocity and likelihood of collection.

Raindrops and ice crystals leaving the cloud loose mass by diffusion. Evaporation and/or sublimation depends on the kinematic viscosity, relative humidity, thermal conductivity, and water-vapor diffusivity. Whether the hydrometeor (rain, ice-crystal, snow, graupel, hail) will reach the ground depends on its size, the temperature and moisture conditions below cloud base, and cloud-base height. Freezing, condensation, and deposition release latent heat thereby warming the ambient air. Evaporation, sublimation, and melting consume heat and cool the ambient air.

Release of latent heat enhances buoyancy and cloud growth, while consumption of heat may initiate downdrafts. Falling hydrometeors may switch a thermally induced upward into a downward motion. The evaporation/sublimation of hydrometeors below cloud base produces negative buoyancy. These processes perturb the temperature, moisture, buoyancy, pressure, and vertical wind field on the cloud-scale.

Within the clouds, both gas-phase and aqueous-phase chemical transformations occur. The degree of solubility characterized by the Henry's law constant affects strongly the distribution of specie between the gas and aqueous phases in clouds. The liquid water and ice content determine the relative importance of gas- and aqueous-phase chemistry in clouds. Freezing of droplets shifts the gas-aqueous-phase equilibrium and thereby modifies the concentration distributions within the vicinity of the freezing particles.

Precipitation removes trace species from the troposphere. Terminal velocities and hence scavenging rates differ as raindrops and ice-crystals differ in shape, size, and density. Evaporation and sublimation below cloud base alter the chemical composition of these layers and of the hydrometeors as the hydrometeor strives to achieve equilibrium with its ambient air.

2.3 Air Chemistry

The atmosphere mainly consists of nitrogen (N_2; 78.08%), oxygen (O_2; 20.95%), Argon (Ar; 0.93%), and CO_2 (0.04%). The noble (non-reactive) and trace gases (reactive) and water vapor account for the rest. Background concentrations differ regionally. Typical midlatitude background concentrations of methane (CH_4), hydrogen oxide (H_2O), and nitrous oxide (N_2O), for instance, are 1.65, 0.58, and 0.33 ppm, respectively. Typical concentrations of carbon monoxide (CO), sulfur dioxide (SO_2), ammonia (NH_3), nitric oxide (NO), nitrogen dioxide (NO_2), ozone (O_3), and nitric acid (HNO_3) are about 0.05–0.2, $10^{-5} - 10^{-4}$, $10^{-4} - 10^{-3}$, $10^{-6} - 10^{-2}$, $10^{-6} - 10^{-2}$, $10^{-2} - 10^{-1}$, and $10^{-5} - 10^{-3}$ ppm, respectively.

2.3.1 Emissions

Many trace gases have natural sources. Natural sources for atmospheric CO_2 and SO_2 are volcanic emissions. Wildfires are natural sources for NO and SO_2. Anaerobic processes in soils and lightning are natural sources of NO. Bacteria decay and decompose organic matter and lead to biogenic emissions of carbon, oxygen, nitrogen, and sulfur. Some emissions are related directly to land-cover/use. Biogenic emissions include, among others, volatile organic compounds (VOC), ammonia and methane. Ammonia emissions are related to agricultural sources and agricultural land-use (e.g., grassland, pasture used for grazing). Rice agriculture and livestock are main sources of atmospheric CH_4. The VOC emissions by forests, grasslands, and swamps probably exceed the anthropogenic VOC emission about six times. Anthropogenic emissions (traffic, combustion for heating and power generation, industrial processes) of trace species occur in connection with urban land-use. CO results from incomplete combustion of carbon-containing fuels.

As these examples illustrate, land-use changes alter the distribution of biogenic and anthropogenic emissions. In the case of some land-cover changes (LCC), the process of the LCC itself leads to emissions. For instance, LCC induced by biomass burning add CO_2 to the atmosphere. The changes in emissions due to LCC alter the atmospheric composition with consequences for air chemistry, the trace gases cycles and, if GHG and particles are involved, the radiation budget.

2.3.2 Background Chemistry

Species emitted directly into the atmosphere such as SO_2, NO, CO and VOC are called primary pollutants (Seinfeld and Pandis 1997). Secondary pollutants result from photochemically processes or chemical reactions of the primary pollutants. Since some of the trace gases are chemically reactive, chemical reactions occur even in a non-polluted atmosphere (background chemistry).

The reaction that triggers the background chemistry is photolysis of ozone at wavelengths of 0.315 µm to form an energetically exited oxygen atom (O^{1D}) and molecular oxygen. In the unpolluted atmosphere, reaction with water vapor produces two highly reactive hydroxyl radicals (OH). Methane reacts with OH to from the methyl radical (CH_3) and water vapor. Hydroxyl radicals also react with CO to form CO_2. Methyl radicals and hydrogen atoms (H) immediately react with O_2 to form methylperoxyl (CH_3O_2) and hydroperoxyl (HO_2) radicals. Peroxy radicals participate in reaction chains that convert NO to NO_2 and produce additional OH and peroxy radical species like methoxyl (CH_3O) or aldehyde (HCHO). Formation of hydrogen peroxide (H_2O_2) or nitric acid (HNO_3) terminates the major reaction chain. Some of the H_2O_2 photolyzes and produces OH. The examination of these chains indicates that CO_2 is the ultimate product of the methane-oxidation chain with CO as a long-lived intermediate state. Formaldehyde (HCHO or H_2CO) formed by reaction of oxygen with CH_3O may photolyze or react with hydroxyl radicals to produce CO and HO_2 radicals. This means formaldehyde is a temporary reservoir of OH (Seinfeld and Pandis 1997).

In the atmospheric boundary layer (ABL), ozonolysis of alkenes can produce OH radicals directly at yields between 7% and 100%, depending on the structure of the involved alkenes. Typically, (organic) peroxyl radicals are co-produced. Compared with the ozone reaction with NO_3 and OH, alkenes-ozone reactions are relatively slow (Seinfeld and Pandis 1997). However, the high concentrations of alkenes in rural areas may offset the low reaction rate.

Since photolysis requires the energy of daylight, different chemical reaction processes occur at night than at day. Consequently, the concentrations of chemical reactive trace gases have a diurnal cycle. We speak of daytime and nighttime chemistry. In Polar Regions, these differences lead to tremendous differences in winter and summer air composition.

2.3.3 Polluted Air

Air will be called polluted air if it contains health adverse substances from human activities or natural processes. In the polluted troposphere, reactive nitrogen compounds (e.g., NO, NO_2, NO_3, N_2O_5, HNO_2, HNO_3) affect strongly the atmospheric chemistry. Ozone initiates most of the primary oxidation chains because it produces OH and nitrate (NO_3) radicals. The O^{1D} produced by ozone photolysis dissipates its excess energy as heat (but not energetically relevant) and recombines with O_2 to build O_3. About 1% of the O^{1D} reacts with water vapor to produce OH radicals. Since the highly reactive OH stems primarily from photochemistry, measurable amounts of OH exist only at day. The average concentration is $\sim 7.7 \cdot 10^{12}$ molecules/m^3 with $5 \cdot 10^{12} - 10 \cdot 10^{12}$ molecules/m^3 in summer and $1 \cdot 10^{12} - 5 \cdot 10^{12}$ molecules/m^3 in winter. Nighttime OH-concentrations typically remain below $2 \cdot 10^{11}$ molecules/m^3.

At night, the nitrate radical is the major reactive oxidant. NO_3 is less reactive than OH, but its nighttime concentrations often exceed the daytime OH-concentrations. NO_3 results from reaction of NO_2 with ozone. At day, NO_3 is quickly photolyzed.

Urban photochemistry does not differ substantially from tropospheric chemistry. However, the range and concentrations of VOC and $NO_x(=NO+NO_2)$ participating in the oxidation permit a greater photochemical turnover than in rural areas. With respect to urban pollution, we distinguish between photochemical smog and London smog.

London-type smog results from burning of high sulfur coal or fuel. In the troposphere, sulfur compounds affect the radiation balance and oxidation of sulfur compounds leads to acidification. Atmospheric oxidation of SO_2 can occur homogeneously as well as heterogeneously in the liquid and gas phases. The gas-phase oxidation leads to formation of sulfuric acid (H_2SO_4). Due to its low vapor pressure, H_2SO_4 can attach quickly to aerosol particles and/or droplets. Most of the H_2SO_4 leaves the atmosphere via wet deposition.

Another loss path for SO_2 is formation of sulfuric acid via Criegee intermediates from ozone-alkenes reactions. The impact of alkene-ozonolysis on OH yields is most likely insignificant. Under high atmospheric moisture conditions, Criegee intermediates are likely to react with H_2O with the major product being hydroxyalkyl hydroperoxide. Removal processes for hydroxyalkyl hydroperoxide are heterogeneous reactions, reactions with OH or photolysis. Photolysis may be a significant source of secondary OH with a hydroxyalkoxy radical (RCH(OH)O) as a coproduct. This radical subsequently may react with O_2 to form a carboxylic acid and HO_2. The removal rates for hydroxyalkyl hydroperoxide are still subject to research (Hasson et al. 2003). The complex aqueous phase reactions of SO_2 depend on droplet size, the availability of oxidants (O_3, H_2O_2) and/or catalysts (Fe, Mn), and the availability of light (Seinfeld and Pandis 1997).

The triad $NO-NO_2-O_3$ is the simplest mechanism of ozone formation in the absence of other reactive species. In the photo-smog of polluted cities, sunlight with wavelength $< 380\,nm$ dissociates tropospheric NO_2 to NO and O, where $O + O_2 + M$ yields $O_3 + M$. Ozone reacts with NO to rebuild NO_2. In absence of any other gases, sources and sinks high (low) concentrations of NO drive the reactions backward (forward).

The triad $NO-NO_2-O_3$ when coupled with peroxy radical-NO reactions enhances ozone formation (Seinfeld and Pandis 1997). Non-methane carbons initiate complex reaction chains that lead to much higher ozone concentrations than possible in the equilibrium of $NO-NO_2-O_3$ alone. In the urban atmosphere, the OH radicals oxidize rapidly a number of other atmospheric pollutants (e.g., NO_2 to HNO3, H_2S to SO_2, SO_2 to H_2SO_4, CH_2O to CO, CO to CO_2). OH radicals may also produce the hydroperoxy radical HO_2 and H_2O_2 radical. Oxidation of O_3 with OH or HO_2 finally destroys ozone and shifts the equilibrium of $NO-NO_2-O_3$. The latter reaction is more important than the former as OH is transferred to HO_2 that primarily reacts with CO and CH_4 and hence provides a link to aqueous chemistry. The reaction of OH and CO may finally result in H_2O_2 formation. Hydrogen peroxide can dissolve in water and oxidize absorbed SO_2 to H_2SO_4.

Starting with the reaction with hydroxyl radicals, methane produces formaldehyde. The reaction of CH_4 with OH produces CH_3O_2. Some of the intermediate products, for instance, CH_3OOH, can dissolve in cloud droplets and/or raindrops. However, usually the concentration of dissolved CH_3OOH is small, and CH_3OOH is not a major oxidant in the aqueous phase. Other reaction chains involve NO_2 as a product. NO_2 can be photolyzed. It can contribute to the formation of tropospheric ozone and the photochemically very reactive HCHO. In tropospheric chemistry, NO_x provides the path for methane oxidation by oxidizing NO to NO_2 and by producing HCHO.

A small quantity of OH oxidizes NO_2 and SO_2 in homogeneous reactions to build weak acids. The reactions with NO_2 are much faster than the reactions with SO_2. Absorption by vegetation and at ocean or other water surfaces and uptake by cloud droplets and raindrops are further sinks of SO_2. This means that LCC may affect tropospheric chemistry due to their impact on the removal process.

2.3.4 Aqueous Chemistry

Sulfate, nitrate, ammonium, and cloud chemical composition vary with time and space due to the natural and anthropogenic sources and meteorological conditions. In the northeastern USA, eastern Canada, East Europe, and Scandinavia, for instance, sulfate dominates the environment because of coal-usage with high sulfur content in urban areas. In the western Unites States, nitrate prevails.

Land-cover changes may affect aqueous chemistry not only because they alter the emission patterns and amounts, but also because they may modify the interception storage; provide water surfaces or remove them; or may affect the frequency of dew, fog or hydrated aerosols, the concentration and type of aerosols and cloud and precipitation formation processes. Water surfaces of any kind namely can absorb atmospheric gases from their immediate vicinity. The dissolved molecules dissociate into ions that may combine with other components and precipitate out as solid particles. Once in solution, atmospheric gases build strong electrolytes (e.g., hydrogen iodide, methane sulfuric acid, nitric acid) that affect the pH-value that is defined as the negative decimal logarithm of the hydrogen-ion activity in solution

$$pH = -\log_{10}[H^+] \tag{2.26}$$

The partitioning of trace species between gas and aqueous phases depends on liquid water content, drop sizes, and the degree of solubility (Henry's law constant). Thus, the relative importance of gas and aqueous-phase chemistry depends on the partitioning of excess water vapor between super-cooled water and ice that may be modified by LCC as discussed in Sect. 2.2. Freezing causes some species to leave the freezing drop as the amount of water decreases and the concentration in the solute increases. Ice chemistry differs from aqueous-phase chemistry and many processes are still unknown. Diffusion of water vapor to or from any water surface and/or collection modify the equilibrium between the gas and aqueous phase. Evaporation of cloud droplets and raindrops is a source for sulfate and nitrate particles. Furthermore, the solution of liquid hydrometeors changes because they scavenge gases from the atmosphere as they settle downward.

In the troposphere, drops are usually far away from Henry's law equilibrium because the mass transfer into the drop may be too slow to reach equilibrium. Observations reveal drop-size dependent compositions of droplets. Some species are enriched in small, others in large droplets. The Henry's law coefficient accounts only for the physical solubility of a given gas. If gases dissolve in a droplet and participate in aqueous reactions, solubility will increase because the effect of acid-base equilibriums is to pull more material into solution (Seinfeld and Pandis 1997). For large drops, the reaction rates can be so high that the mass transport into the drop limits the reactions. Reactions in drops with high pH-values (e.g., radiation fog pH $\sim$ 7) are also mass-transport limited. Typically, large drops are less acidic than small drops.

Common aqueous reactions start with the diffusion of SO_2 to a droplet/water surface, and its dissolution therein. In the solution, various heterogeneous reactions convert the dissolved SO_2 to sulfate ions. The dissolved SO_2 forms sulfurous acid (H_2SO_3), bisulfite ions (HSO_3^-), and sulfite ions (SO_3^{2-}). Conversion to sulfuric acid (H_2SO_4), bisulfate ions (HSO_4^-), and sulfate ions (SO_4^{2-}) leads to acid rain. Over a broad range of pH-values, oxidation with H_2O_2, followed by O_3, some metal-catalytic reactions, and NO_2 are the most important reaction paths in cloud droplets and raindrops. Catalytic reactions with OH, Cl_2^-, and Br_2^+ also occur. While most reactants available in the drops stem from the gas phase, H_2O_2, HO_2, and OH can also be produced photochemically in the solution. Raindrops remove the sulfuric acid and bisulfite (HSO_3) from the atmosphere (Seinfeld and Pandis 1997).

Oxygen dissolves easily in water, but oxidizes slowly therein unless catalytic processes occur. Depending on the pH-value, catalytic processes involving iron (Fe^{3+}) and manganese (Mn^{2+}) increase significantly the reaction rate. Iron and manganese exist in traces in drops. Coexistance of iron and manganese can enhance the catalytic rates three to ten times the sum of the single catalytic rates, and the catalyzed reaction of O_2 may exceed the oxidation of O_3 (Seinfeld and Pandis 1997).

At pH > 5 or so, O_3 oxidation exceeds that of H_2O_2 that otherwise has the highest reaction rates. If H_2O_2 is the main oxidant, sulfate production will be less sensitive to the pH-value and liquid water content.

In the sulfate production, the oxidation path via O_3 becomes more important in raindrops below cloud base and close to the cloud top of not entirely glaciated clouds. Except for very low pH-values, reactions with O_3 are quicker than reactions with NO_2. Ammonia is a net source of alkalinity in the atmosphere. Its reaction with sulfuric and nitric acids produces aerosols (Seinfeld and Pandis 1997).

The oxidation rates depend on water temperature. As temperature decreases, the reaction constants decrease and solubility increases, but the latter effect is more efficient than the former. This means that LCC-induced temperature changes may affect aqueous chemistry.

2.3.5 Gas-to-Particle Conversion

Gas-to-particle conversion refers to both the condensation of gases to form new particles or onto other particles. Gas-to-particle conversion provides more particles than direct emissions. Condensation on particles prevails for high surface area of existing particles and gases with low supersaturation. The new particles are of Aitken-nucleus size and may serve as CCN. Particle growth influences the subsequent CCN activity, visibility, and climate.

Sulfur, nitrogen, organic and carbonaceous materials are the major chemical species involved in gas-to-particle conversion. Volatile organic compounds and gases from various plants and bacteria may act as precursors in the gas-to-particle-conversion of organic and carbonaceous particles. Since LCC changes alter the biogenic emissions, they may affect gas-to-particle conversions indirectly with further consequences for the energy and water cycles.

2.3.6 Removal of Gases and Aerosols

Gaseous species are removed from the atmosphere by wet or dry deposition. Dry deposition depends on the height above the surface, particular specie being deposited, meteorological conditions in the ASL, and the surface characteristics, including soil type and plant species (e.g., Kramm et al. 1995). The dry deposition process can be treated by a resistance network analog to Ohm's law and Kirchhoff's law of electrostatics,

$$F = -v_{\mathrm{d}} C_{\mathrm{R}} \tag{2.27}$$

Here, C_{R} is the partial density of the species and

$$v_{\mathrm{d}} = \frac{1}{r_{\mathrm{a}} + r_{\mathrm{s}} + r_{\mathrm{t}}} \tag{2.28}$$

is the dry deposition velocity. Furthermore, r_{a}, r_{s}, and r_{t} are the aerodynamic resistance, sorption resistance (often called the surface resistance), and transfer resistance, respectively. The aerodynamic resistance describes the turbulent diffusion of constituents from the free atmosphere to the surface of the laminar surface layer. It depends on surface and meteorological conditions (e.g., surface roughness, wind, atmospheric stability). The surface-layer resistance is a function of the molecular transfer. The transfer resistance describes the physicochemical interaction between the specie and surface. Thus, dry deposition not only depends on the micrometeorological conditions but also on land-cover/use, soil conditions, and soil type. Vegetative controls (e.g., the sensitivity to PAR, specific humidity deficit between leaf and ambient air, foliage temperature, soil-moisture deficit, volumetric CO_2 concentration) affect the stomatal resistance for uptake of air. These dependencies of dry deposition on vegetation and land-surface conditions means that LCC may alter the residence time of gases in the atmosphere and the distribution of dry deposition.

Exhalation of trace species (e.g., NO, NO_2, CH_4) from land surfaces may occur. This exhalation can be considered by inserting Kramm's μ-factor into Eq. 2.28, where the numerator has to be replaced by $1 - \mu$. This factor allows to distinguish between deposition ($0 \leqslant \mu < 1$), compensation ($\mu = 1$), and emission ($\mu > 1$). It may be expressed as empirical functions of time (Kramm and Dlugi 1994).

The flux-resistant approach that is based on the constant flux approximation seems to be appropriate to provide reasonable flux results if the species are chemically conservative (i.e., the response time t_{c} of the chemical processes is much larger than the response time t_{d} of the diffusive transfer processes; $t_{\mathrm{c}} \gg t_{\mathrm{d}}$) or always in equilibrium with their reactants ($t_{\mathrm{d}} \gg t_{\mathrm{c}}$). Thus, in order to fulfill the constant flux requirement, the response times t_{c} and t_{d} must differ considerably from each other (Kramm and Dlugi 1994). Since the chemical reaction-rate constants depend frequently on the properties of the air (e.g., temperature, pressure) and the turbulent mixing of the air is affected strongly by the thermal stratification, the ratio of the response times $N_{\mathrm{D}} = t_{\mathrm{d}}/t_{\mathrm{c}}$ – denoted as Damköhler number – can change within a wide range of mutual influences (see Fig. 2.7). The application of the constant flux approach

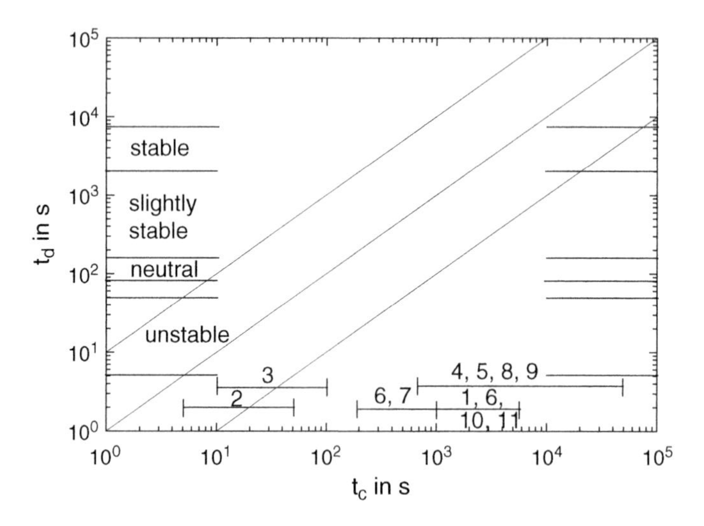

Figure 2.7. Relationship between the characteristic response times of the chemical processes t_c, and the diffusive transfer processes t_d, for different chemical reactions in chemical cycles for a surface layer thickness of $\Delta z = 10\,\mathrm{m}$ and different thermal stratification conditions (Modified after Kramm and Dlugi 1994)

may also provide inaccurate results if unaccounted for emissions into the layer occur (Kramm and Dlugi 1994).

Dry removal of aerosols from the atmosphere includes sedimentation, coagulation, and impaction. Sedimentation removes about 10–20% of the aerosol mass. The deposition velocity of aerosols is given by

$$v_\mathrm{d} = -\frac{v_\mathrm{T}\,(1-\mu_\mathrm{p})}{1-\exp\left(-v_\mathrm{T} r_\mathrm{a}\right)} \tag{2.29}$$

where for a given aerosol spectrum v_T is the terminal velocity and μ_p is Kramm's μ-factor, and $r_\mathrm{a} = r_\mathrm{mt} + r_\mathrm{t}$.

The scavenging rates depend on the characteristics of the scavenging hydrometeors (e.g., shape, size, density, terminal velocities). The washout ratio is the ratio of the concentrations of the respective gases or particles in the surface-level precipitation to the concentration of the respective specie in surface-level air. The sum of wet removal from the column aloft determines the wet deposition onto a land surface preassuming that the scavenged pollutants remain in the hydrometeors. The wet deposition flux depends on the aqueous concentration and precipitation intensity. Since typically small (large) droplets contain more sulfate (nitrate), wet deposition varies with particle size and specie.

1. $HO_2 + HO_2 \rightarrow H_2O_2$
2. $HNO_3 + NH_3 \rightarrow NH_4NO_3$
3. $O_3 + NO \rightarrow NO_2 + O_2$
4. O_3 + Isoprenes → Products (P)
5. O_3 + Monoterpenes → P
6. NO_3 + Monoterpenes → P
7. NO_3 + Isoprenes → P
8. OH + Isoprenes → P
9. OH + Monoterpenes → P
10. O_3 + Olefins → P
11. $O_3 + NO_2 \rightarrow NO_3 + O_2$

Strong vertical gradients in ammonium concentration may exist due to the distribution of sources.

Various land surfaces may scavenge droplets and/or ice crystals from fog or clouds encountering them. As the droplets hit the surface (e.g., forest on a mountain barrier), the dissolved trace gases, particles, and the reaction products in the droplets are deposited. Since, on average, the acidity of small droplets is higher than of large drops, occult deposition may be a greater burden for ecosystems than acid rain.

2.4 Interaction Between Energy, Water, and Trace Gas Cycle

The energy, water, and trace gas cycles are closely linked with land-cover. Biogeochemical cycle refers to a cyclic system, in which chemical elements transfer between biotic and abiotic parts of ecosystems. The distributions of temperature, water, and liquid water content control the biogeochemical cycles, and affect the sources and sinks of the gases. Important biogeochemical cycles affected by LCC are the energy, hydrological, carbon, nitrogen, oxygen, and sulfur cycles. Since plants use CO_2 for photosynthesis, any LCC modify the carbon, energy, and water cycles. Changes in the atmospheric CO_2 concentration may shift the equilibrium of the carbon cycle with potential consequences for other biogeochemical cycles. Consequently, assessment of the long-term impacts of LCC on the atmosphere may have biogeochemical aspects.

Changes in agricultural practices (e.g., use of organic or nonorganic fertilizer, burning of the field) or land-cover/use affect the exchange of nitrate components between the surface and the atmosphere. Plant decay and animal waste add nitrates to the soils where bacteria and fungi mineralize organic N to NH_4^+. Bacterica use nitrogen as energy source for chemical reactions (similar to the respiration in other organisms) and produce NO and N_2O when converting NH_4^+ to NO_3^- – a process called nitrification. Vegetation takes up nitrates from the soil. This N_2 to organic N transfer is called N-fixation. Uptake by plants converts NO_3^- and NH_4^+ to organic N during photosynthesis. Bacteria form

NO_3^- to N_2 thereby producing N_2O. Dinitrification reduces nitrates in the soil to molecular nitrogen NO. Both NO and NO_2 are released to the atmosphere.

Changes in graupel formation affect the frequency of lightning and wildfire-induced LCC. As lightening flashes produce nitrogen oxides, any changes in the frequency and distribution of graupel formation modify the distribution of nitrogen oxides and species reacting therewith. Changes in wildfire frequency and distribution affect the carbon cycle.

References

Avissar R (1993) Observations of leaf stomatal conductance at the canopy scale: an atmospheric modeling perspective. Bound Layer Meteorol 64:127–148

Bonan GB (2008) Forests and climate change: forcings, feedbacks, and the climate benefits of forests. Science 320:1444–1449

Chen F, Dudhia J (2000) Coupling an advanced land-surface/hydrology model with the Penn State/NCAR MM5 modeling system. Part I: Model description and implementation. Mon Wea Rev 129:569–585

Claussen M (1990) Area-averaging of surface fluxes in a neutrally stratified, horizontally inhomogeneous atmospheric boundary layer. Part A General topics. Atmos Environ 24:1349–1360

Damköhler G (1940) Influence of turbulence on the velocity of flames in gas mixtures. Z Elektrochem 46:601–626

Deardorff JW (1978) Efficient prediction of ground surface temperature and moisture, with inclusion of a layer of vegetation. J Geophys Res 83C:1889–1903

Dingman SL (1994) Physical hydrology. Macmillan, New York.

Dolman AJ (1992) A note on areally-averaged evaporation and the value of the effective surface conductance. J Hydrol 138:583–589

Giorgi F, Avissar R (1997) Representation of heterogeneity effects in earth system modeling: experience from land surface modeling. Rev Geophys 35:413–438

Hasson AS, Chung MY, Kuwata KT, Converse AD, Krohn D, Paulson SE (2003) Reaction of Criegee intermediates with water vapor: an additional source of OH radicals in alkene ozonolysis? J Phys Chem A 107:6176–6182. doi:10.1021/jp0346007

Jarvis PG (1976) The interpretation of the variations in leaf water potential and stomal conductance found in canopies in the field. Philos Trans R Soc Lond 273B:593–610

Koster RD, Suarez MJ (1992) A comparative analysis of two land surface heterogeneity representations. J Clim 5:1379–1390

Kramm G, Dlugi R (1994) Modelling of the vertical fluxes of nitric acid, ammonia, and ammonium nitrate in the atmospheric surface layer. J Atmos Chem 18:319–357

Kramm G, Dlugi R, Dollard GJ, Mölders N, Müller H, Seiler W, Sievering H (1995) On the dry deposition of ozone and reactive nitrogen compounds. Atmos Environ 29:3209–3231

Kramm G, Beier N, Foken T, Müller H, Schröder P, Seiler W (1996) A SVAT scheme for NO, NO_2, and O_3- model description. Meteorol Atmos Phys 61:89–106

Lhomme J-P (1992) Energy balance of heterogeneous terrain: averaging the controlling parameters. Agric For Meteorol 61:11–21

Mahrt L, Sun J (1995) Dependence of exchange coefficients on averaging scale and grid size. Q J R Meteorol Soc 121:1835–1852

Mölders N (1999) Einfache und akkumulierte Landnutzungsänderungen und ihre Auswirkungen auf Evapotranspiration, Wolken- und Niederschlagsbildung. University of Leipzig, Leipzig

Mölders N (2001) On the uncertainty in mesoscale modeling caused by surface parameters. Meteorol Atmos Phys 76:119–141

Mölders N, Raabe A (1996) Numerical investigations on the influence of subgrid-scale surface heterogeneity on evapotranspiration and cloud processes. J Appl Meteorol 35:782–795

Mölders N, Raabe A, Tetzlaff G (1996) A comparison of two strategies on land surface heterogeneity used in a mesoscale β meteorological model. Tellus 48A:733–749

Mölders N, Haferkorn U, Döring J, Kramm G (2003) Long-term numerical investigations on the water budget quantities predicted by the hydro-thermodynamic soil vegetation scheme (HTSVS) – Part II: Evaluation, sensitivity, and uncertainty. Meteorol Atmos Phys 84:137–156

Mölders N, Jankov M, Kramm G (2005) Application of Gaussian error propagation principles for theoretical assessment of model uncertainty in simulated soil processes caused by thermal and hydraulic parameters. J Hydromet 6:1045–1062

Pielke RA (2002) Mesoscale meteorological modeling. Academic, New York

Rosenberg NJ, McKenney MS, Martin P (1990) Evapotranspiration in a greenhouse-warmed world: a review and a simulation. Agric For Meteorol 47:303–320

Seinfeld JH, Pandis SN (1997) Atmospheric chemistry and physics, from air pollution to climate change. Wiley, New York

Seth A, Giorgi F, Dickinson RE (1994) Simulating fluxes from heterogeneous land surfaces: explicit subgrid method employing the biosphere-atmosphere transfer scheme (BATS). J Geophys Res 99:18651–18667

Wiscombe WJ, Warren SG (1980) A model for the spectral albedo of snow. I: Pure snow. J Atmos Sci 37:2712–2733

Chapter 3

Impact of Land-Cover and Land-Cover Changes

3.1 Detection of Land-Cover Changes

When the first satellite imagery became available, they evidenced human impacts on land-cover on a large scale. They showed cities, deforestation pattern, international borders, and even land-cover changes (LCC) and their impacts from space (Fig. 3.1). The Negev, for instance, appears much darker than the Sinai where grazing is unrestricted (Nicholson et al. 1998). Clouds frequently form over the dark native vegetation that is separated from the agriculturally used area by the more than 3,258 km long rabbit fence in Australia.

Land-cover is defined by the vegetation, hydrologic, and geologic features and man-made structures of the land surface. Most land-cover classifications characterize an area of defined size as a discrete land-cover type. Discrete classes do not necessarily reflect the gradients or mosaics of the land-cover types across the landscape. Consequently, the representation of a landscape by any land-cover dataset depends on the resolution, the number of land-cover types distinguished, the data sources, and the particular method used to derive the land-cover classification. If, for instance, a land-cover classification derived from satellite data is to have a scale in the range 1:25,000–1:50,000, the spatial resolution of the sensors requires a pixel size of about 15–30 m.

Land-cover datasets derived at the same time, but from different sources may differ drastically (e.g., Fig. 3.2). In the case of satellite data,

N. Mölders, *Land-Use and Land-Cover Changes*, Atmospheric and Oceanographic Sciences Library 44, DOI 10.1007/978-94-007-1527-1_3,

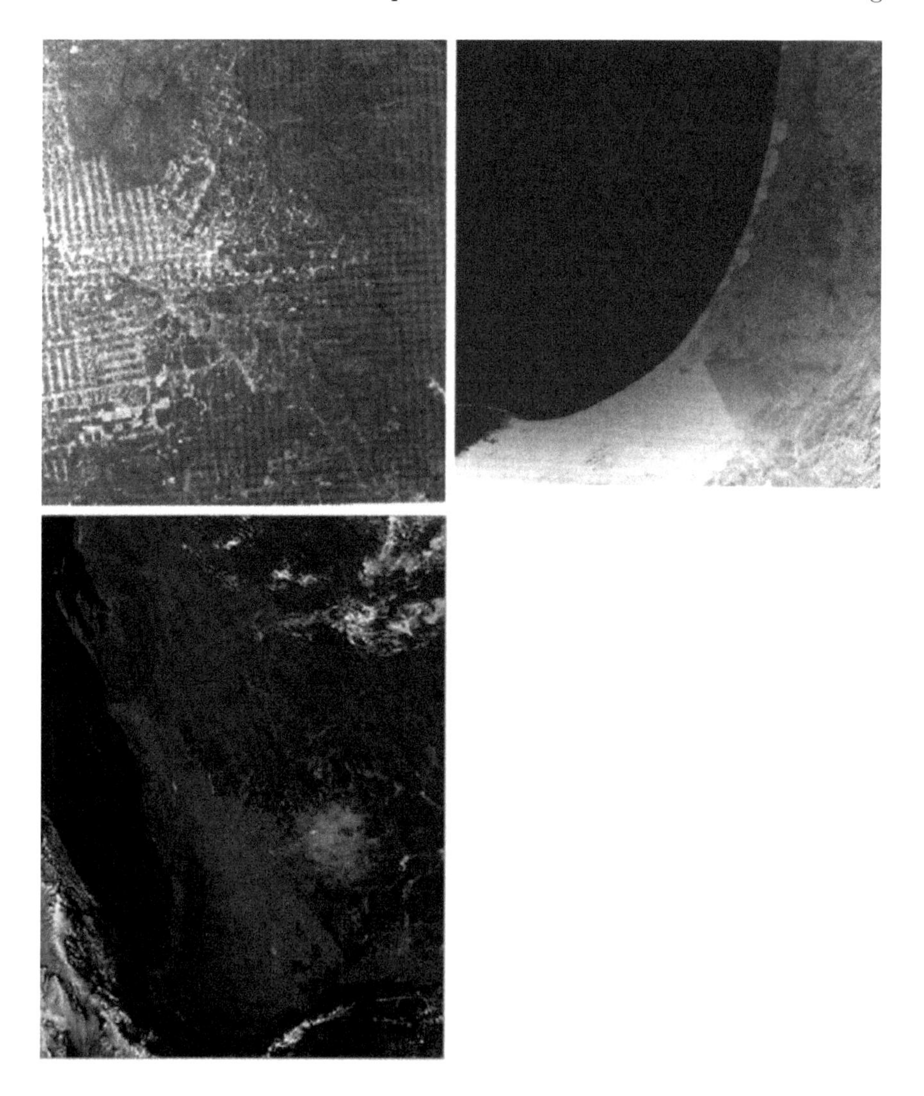

Figure 3.1. Satellite imagery showing the low and high albedo forest and clear-cuttings in the Amazon (*top left*), the demarcation line between the low and high albedo of the Negev and Sinai (*right*), and the high and low albedo of agricultural and native vegetation, respectively, and clouds over the native vegetation (*bottom left*) (Upper images from Cotton and Pielke (2007), lower image modified from `http://www.nytimes.com/2007/08/14/science/earth/14fenc.html`)

such differences may be due to the resolution of the pixels. The pixels of the Advanced Very High Resolution Radiometer (AVHRR), for instance, have a resolution of about 1.1 km at the subsatellite point, while that of the High Resolution Visible (HRV) instrument is approximately 20 m and permits to resolve single fields.

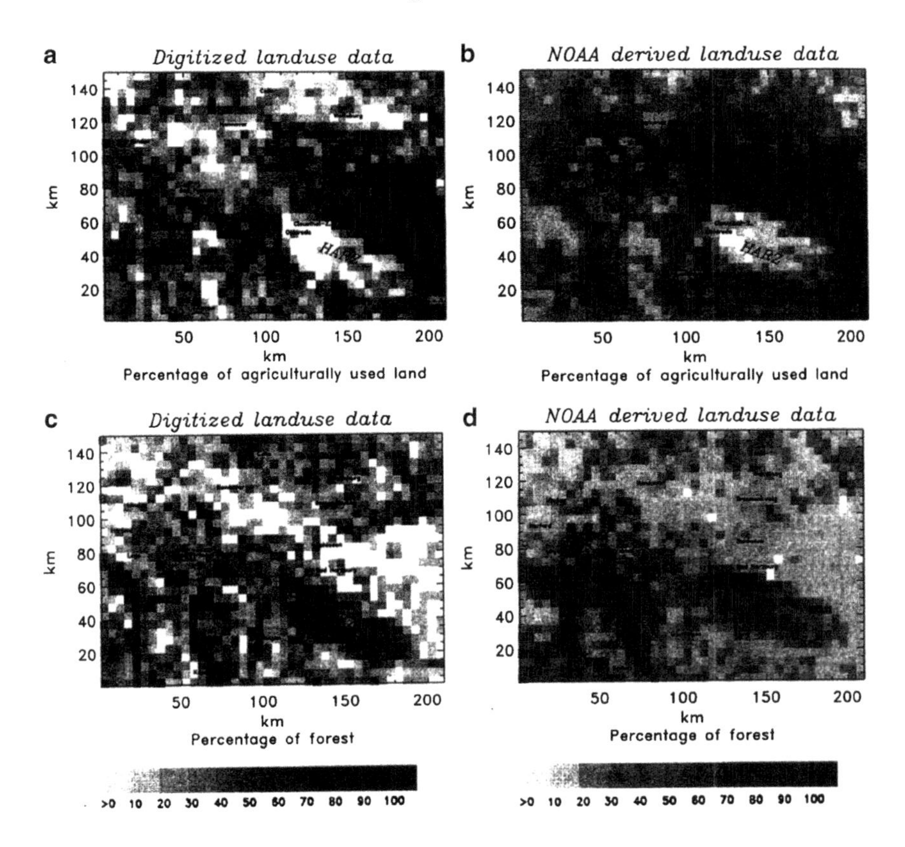

Figure 3.2. Percentage of agriculturally used land (**a**, **b**) and forest (**c**, **d**) within each $5 \times 5\,\text{km}^2$ area as derived from digitized maps (*left*) and AVHRR data (*right*) for an area in central Germany. Areas wherein the respective land-cover type does not occur are white. Note that differences for urban land and grassland are of similar order of magnitude (Modified after Mölders et al. (1997))

The temporal resolution affects what information can be derived. The high temporal resolution of the AVHRR, for instance, permits observing the phenological vegetation changes during the annual cycle, while the low temporal, but high spatial resolution of the HRV permits mapping year-to-year crops changes on fields.

3.1.1 Land-Cover Maps

Prior to the satellite era, map production and land-cover classification relied on geodetic surveys. Thus, land-cover data representing landscapes prior to the satellite era typically base on digitized maps of a

1:100,000 scale or finer. Most of these maps distinguish less than ten land-cover types. Often these maps fail to distinguish explicitly between grassland and agriculture, or between different forest types, or classify water meadows as grassland. Such and similar shortcomings will underestimate the extension of certain land-cover types (e.g., the fractional coverage by grassland far away from rivers). Often the classification of land-cover or number of land-cover types distinguished differs from one edition to the next. Such changes and the classification practice may hide later LCC and/or may pretend LCC that actually did not occur.

Typically, digitizing procedures count the pixels of each land-cover type within a squared area. The length of the square determines the resolution of the resulting land-cover dataset. In that area, n different scenes, namely, $n-1$ land-cover types plus the pixels of the digitizing line may occur. Digitizing procedures then aggregate the pixels to the dominant land-cover type within the area. If the number of pixels of two land-cover types is equal, but exceeds that of all other land-cover types within that square area, typically the digitizing lines will be added to the land-cover type lying on the concave side to determine the dominant land-cover. All non-dominating land-cover types are of subgrid-scale with respect to the resolution of the digitizing grid. The resolution of the digitizing device (pen) and the map limit the resolution of the digitized data.

Some digitized land-cover datasets provide error estimates. Such estimates can be derived by analyzing the typical length of patches of equal land-cover on maps with higher resolution (e.g., 1:25,000 scale maps when the digitized data base on 1:100,000 scale maps). The error of digitized land-cover data depends on the number of land-cover types distinguished and the chosen resolution. Assuming that the patches are squares and weighting with the frequency of occurrence provides the average error for a $1 \times 1\,\mathrm{km}^2$ area for the various land-cover types distinguished. If, for instance, a 1-km resolution is aimed for and nine land-cover types are distinguished, the error can be as large as 89% if all land-cover types and the digitizing lines occur within the $1 \times 1\,\mathrm{km}^2$ area. The probability of this case, however, is very low. As this examples shows, digitizing yields greater errors for heterogeneous than predominantly homogenous landscapes.

Discrepancies in digitized data may result from information loss due to the digitizing procedure, subjective decisions by the digitizer due to not well-defined land-cover boundaries, and the accuracy and details of the maps that differ among land-registry offices. Dissimilar definitions of ecosystems and/or national differences in required stand density for classification as forest lead to differences in the description of the same landscape on both sides of the borders.

3.1.2 Remote Sensing–Derived Land-Cover

Since the spectral responses of different land-cover differ at different wavelengths, land-cover characteristics can be classified by means of multispectral remote sensing data. Land-cover distribution maps have been derived from various satellite-based radiometers and instruments (e.g., Landsat Thematic Mapper [TM], Satellite pour l'Observation de la Terre [SPOT] HRV, Aqua and Terra Moderate Resolution Imaging Spectroradiometer [MODIS], Polar Operational Environmental Satellite [POES] AVHRR, Synthetic Aperture Radar [SAR]).

Coarse-resolution data with pixels >250 m (e.g., AVHRR, MODIS) have been used to obtain land-cover information over large regions. Medium-resolution data with pixels of 5–80 m across (e.g., Landsat Multi-Spectral Scanner (MSS), TM) have been applied successfully for general mapping efforts in urban/suburban and natural environments. High-spatial resolution sensors with pixel sizes <5 m have been used on airborne systems. Data from these sensors have been applied for the investigation in both urban/suburban and natural landscapes. The major advantage of airborne systems is the independency from satellite overpasses.

The primary challenge in deriving accurate land-cover information represents the general remote sensing problem of the maximization of the signal-to-noise ratio. Inherent noise limits the land-cover classification. Preprocessing is required for atmospheric correction, image registration, geometric correction, mosaicing, sub-setting and masking of contaminated (e.g., clouds, shadows) and/or irrelevant scene fragments. Often, additional data like the Normalized Difference Vegetation Index (NDVI) are calculated using spectral radiances of different channels. The NDVI, for instance, permits to distinguish between non-vegetated and vegetated areas due to the strong positive relationship between upwelling radiance and vegetation.

Satellite-derived land-cover data is very sensitive to the classification algorithm used. Any land-cover classification algorithm makes use of the unique combination of spectral responses of the various land-cover types. These algorithms may consider hard, discrete categories and assign each pixel to one class only or fuzzy categories and assign a pixel proportionally to multiple classes. Some algorithms process the entire image pixel-by-pixel, while other algorithms decompose the image into homogeneous patches for object-oriented classification.

Commonly applied land-cover classification algorithms reach from visual interpretation or multivariate statistical methods based on multispectral and textural differences using data from a single satellite overpass to classifications considering the spectral signature of the underlying surface during several overpasses.

The earlier studies of land-cover classification used simple thresholding (visual or statistically based) in multiple spectral bands and pattern-recognition procedures such as edge enhancement, segmentation and spatial analysis rules, or maximum likelihood classifiers. Principal component analysis (PCA) or multivariate alteration detection (MAD) apply data transformation. In these methods, the data of the n spectral bands per image are transferred into the n-dimensional space. Typically, all desired bands and/or derived quantities (e.g., NDVI) are composited into a layer stack. An unsupervised or supervised classification using calibration data is performed on the dataset to classify the land-cover based on all data in the stack. Some modern studies have used artificial intelligence techniques both supervised and unsupervised, or ensemble-classification methods for a land-cover classification. Many of these remote sensing classification techniques assume that the data for a land-cover class follow a multivariate Gaussian distribution. More advanced techniques are the Independent Component Analysis (ICA) mixture model, Support Vector Machines (SVM), Markov Random Field (MRF) models, and the spectral unmixing procedure.

The ICA mixture model considers the radiometer data as a mixture of several mutually exclusive classes that can be described by a linear combination of independent sources with non-Gaussian probability-density functions. The ICA mixture model uses an extended information maximization-learning algorithm to identify independent components and the mixing matrix for each land-cover class. It also delivers for each pixel the probability that it belongs to the identified land-cover class.

MRF models are able to consider local interactions among neighboring pixels. Thus, they can be used to explain various characteristics existing among neighboring pixels. Since changes are more likely to exist in adjacent pixels rather than at disjoint points, MRF models have been developed that exploit the statistical correlation of intensity values among neighboring pixels. In coarse resolution remote sensing images, generally most pixels have spectral responses dominated by more than one land-cover class (mixed pixels). In general, mixed pixels degrade the quality of any land-cover classification. MRF models are able to capture this sub-pixel information from coarse resolution images.

SVM rely on statistical learning theory. Typically, an image is segmented initially into various objects containing more than 100 pixels. These objects are identified for training to distinguish the given number of land-cover classes. After the training phase, the obtained SVM algorithm classifies the entire image.

The increase of spectral resolution led to the development of spectral mixture analysis (SMA). This procedure assumes that image elements are composed of multiple pure spectral signatures called endmembers. This assumption means that the measured spectrum of a given pixel is a linear superposition of the reflectance properties of the pixel components. For a given spectrum and given components of the pixel, the fractional contribution of each endmember to the measured spectrum can be determined from superposition of the endmembers and minimizing the error between the measured and calculated spectrum. Since the SMA method assumes that the reflective properties add up linearly, the fractional contribution of the reflectance of each endmember to the total spectrum equals the fractional coverage of the endmember. Thus, the spectral unmixing procedure will provide the percent coverage of the various land-cover types in each pixel if the reflectance properties of the individual land-cover types are defined. This means the spectral unmixing procedure determines land-cover at the sub-pixel scale.

Land-cover maps derived from satellite imagery are usually imported into a Geographic Information System (GIS). Each pixel is treated as an independent observation independent of its spatial location. A major shortcoming is that an area may contain some pixels that belong to another class. In such case, extensive editing is necessary before the maps can be imported into a GIS. Alternatively, image segmentation can be applied. Herein clusters of adjacent pixels that represent a meaningful class from a user's point of view are combined. In a next step, the sets of objects are classified according to their average responses in each of the spectral bands.

Misclassification leads to discrepancies among land-cover datasets. In the case of satellite-derived land-cover classifications, the area represented by a pixel at the edges of the passage is the largest due to the Earth's curvature and the geometric conditions of the system Earth-satellite. Thus, at the edges of a satellite scan along its passage, the likelihood that more than one land-cover type is in the area represented by the pixel increases and enhances the risk of classification errors. Other error sources are calibration errors, shadows of mountains, and the direction dependency of the solar reflection and the terrestrial emittance.

3.1.3 Land-Cover Change Detection

Land-cover changes are dynamic processes and occur at varying rates and in different locations in response to environmental, economic, and social factors. The rate of change can be gradual, dramatic, subtle, or abrupt. Wildfires, for instance, yield abrupt changes, while biomass accumulation or migration of ecosystems occurs gradually. Thus, LCC can be considered as a continuum or a categorical variable (class).

Various methods exit to identify LCC. Change in land-cover can be identified by visual interpretation or digital detection techniques. Visual analysis is difficult to replicate. Individual interpreters provide different results. Since many of the radiometers have been providing data for nearly 20 years, data taken over the same area, but several time apart can be used to detect and document LCC.

GIS technology can be used to deliver land-cover change maps, derived from any two digitally available land-cover datasets. These digital maps can stem from the methods described in the previous section, but preferably should have been derived the same way.

Digital change detection permits the quantification of temporal phenomena from multi-date imagery of multispectral sensors. The sensors have to be identical and calibrated so data from different years are comparable. Change detection between pairs of land-cover data (bitemporal) as well as between time-profiles of land-cover derived indicators (temporal trajectories) can be made (Coppin et al. 2004).

Many algorithms to classify LCC from remote sensing data have been developed. Most digital change-detection methods use per-pixel classifiers and pixel-based change information. Most of them combine change extraction (change-detection algorithm) and change separation/labeling (change-classification routine). Knowledge on potential changes serves in detecting LCC in the separation/labeling procedure. These separation/labeling rules are independent of the detection algorithm, but depend on the kind of LCC.

Like land-cover detection algorithms, all LCC-detection algorithms have to maximize the signal-to-noise ratio. In addition, they face the noise from differences in atmospheric absorption and scattering caused by interannual variability of water vapor and aerosol concentrations, aerosols from major volcanic eruptions, and sensor calibration inconsistencies. Another challenge is that LCC-detection algorithms must be capable to deal adequately with the initial static situation, but also interpret the LCC. This means they must account for variability at the seasonal scale. Furthermore, LCC-detection algorithms must identify the spatial extent and the context of the LCC, and the change classes should be mutually exclusive.

The kind of LCC to be detected affects the philosophy of the developed/chosen algorithm. The temporal sampling rates must match the intrinsic scales of the processes to be monitored (Coppin et al. 2004). Data acquisition and algorithms for detecting floods or biomass burning, for instance, differ from those that serve to document deforestation or urbanization.

The bitemporal change detection requires overall comparable phenological conditions (e.g., two peak-green summer images). Bitemporal change detection uses anniversary windows to minimize discrepancies in reflectance caused by seasonal sun angle and vegetation-growth differences. These anniversary windows must be several years apart to detect LCC. The type of LCC to be detected determines how many years are required at minimum. The main challenges of the bitemporal change detection are the calendar acquisition dates and the change-interval length (temporal resolution). Major inaccuracy of bitemporal change detection results from phenological disparities due to interannual and spatial variability of temperature and precipitation. In most regions, bitemporal change detection works best in summer and winter because of their phenological stability (Coppin et al. 2004).

The temporal trajectory analysis relies on the distinct seasonal trends of the various land-cover types. This method needs data on a continuous timescale between growing seasons with similar meteorological conditions. Low-orbiting satellites (e.g., Aqua, Terra, POES) can provide the high temporal resolution required by this method. The coarse spatial resolution of the radiometers on board of these satellites, however, limits the LCC categories and the size of LCC that can be identified. The major advantage of the temporal trajectory analysis is that it eliminates phenological influences, because it separates seasonal changes from other changes (Coppin et al. 2004).

The most widely applied change-detection algorithm is univariate image differencing. This method provides information where changes occurred, but no information on the kind of changes. First, imageries of two different dates are precisely registered. Then the original or transformed (e.g., NDVI, albedo) imagery of the first date is substracted from the imagery of the later date. The resulting matrix of positive and negative values represents areas of change, while zeros indicate no change (Coppin et al. 2004).

Post-classification comparison also called delta classification use spectral classification results of different dates that are produced independently. The classification results are compared on a pixel-by-pixel or segment-by-segment basis to derive a matrix of LCC and LCC classes.

The major advantage of delta classification is that the separate classification minimizes the problems from radiometric calibration between dates. The main disadvantage is that the accuracy of LCC-detection depends on the accuracy of the initial classifications. The challenge is to ensure consistent, analogous, and highly accurate target identifications (Coppin et al. 2004).

Composite analysis (e.g., spectral/temporal change classification, multi-date clustering, spectral change pattern analysis) requires prior knowledge of the logical interrelationships of LCC classes. Composite analysis applies combined registered datasets or corresponding subsets of bands, collected under similar phenological conditions, but in different years. The method relies on identifying classes with LCC from those without LCC by their statistically significant differences. In principle, composite analysis only requires a single, but very complex classification when a discriminant analysis was applied prior to application (Coppin et al. 2004).

When applied for LCC detection, PCA or MAD use bitemporal linear data transformation. Herein the data of n spectral bands per image of two-date imageries are transferred into the $2n$-dimensional space. The major fraction of the variance can be associated with correlated (constant, unchanged) land-cover. Areas that experienced LCC are enhanced in the lower components.

Other algorithms applied for LCC detection are the SMA and SVM. Within the framework of LCC detection, the idea is that LCC modify the end-member proportions and hence give the fractional changes in land-cover classes within a pixel or the imagery.

3.1.4 Role of Land-Cover Data in Land Cover–Related Studies

Numerical weather prediction (NWP), air quality, and climate models require, among other things, information on land-cover as lower boundary conditions to describe the energy, water, and trace gas cycle. Due to the exchange of heat, momentum, and matter at the land-atmosphere interface, the quality of the land-cover dataset may affect the model results. In many climate studies, often the response of land-cover to changing climate is of interest and has to be simulated by dynamic ecosystem models. In NWP, land-cover data and their accuracy are essential for local scale forecasts.

Various authors examined the impact of land-cover data on the predicted variables of state, water, and energy fluxes. Mölders et al. (1997), for instance, performed simulations with a mesoscale model once with digitized maps and once with satellite-derived land-cover data for the Harz region, Germany (Fig. 3.2). Despite the fractional coverage of the various land-cover types varied only within ±5%, the location of land-cover types differed appreciably. In areas of large discrepancies in land-cover, daily mean near-surface temperatures, sensible and latent heat fluxes differed up to 2.3 K, 30 W m^{-2} (29%), and 32 W m^{-2} (34%), respectively, while the regional averages of these quantities changed less than ±2%. If in an area, low vegetation was classified as settlements or forest, or vice versa, simulated fluxes would differ the greatest. In regions dominated by the same land-cover in both datasets, area-averaged fluxes differed up to 165 W m^{-2} (35%), except for grassland during daytime for which differences were the smallest.

In air-quality modeling, land-cover data are not only critical to simulate the meteorology correctly, but also for the calculation of biogenic emissions and dry deposition of gases and aerosols. Thus, inaccurate land-cover data may have huge impacts on simulated air quality and input into ecosystems. Gulden et al. (2008), for instance, evaluated the sensitivity of biogenic emissions to land-cover datasets on a 0.1°-grid over Texas, USA, from 1993 to 1998. They applied a satellite-derived and a survey-derived land-cover dataset to simulate biogenic emissions. Systematic variation of the datasets showed that the representation of bare-soil fraction, vegetation-type distribution, and phenology causes differences as large as a factor of three in simulated mean statewide total biogenic emissions. Biogenic emissions seem to change nearly linearly in response to leaf-area index (LAI; one-sided green leaf area per unit ground area) changes. Simulated biogenic emissions differed the strongest in hotspots of biogenic emissions and for conurbations. Unfortunately, in conurbations, tropospheric ozone poses an air-quality issue, and emissions may modify the ozone concentrations strongly. The authors reported that statewide average biogenic emissions differ by a factor of 1.7 due to uncertainty in bare-soil fraction. The fraction of bare soil affects strongly the vegetation temperature and hence indirectly the biogenic emissions. Similar relations between land-cover and biogenic emissions were found for other regions.

3.2 Observations and Major Field Experiments

Unfortunately, historic data documenting the impact of LCC on local weather and climate are rare. Most documented cases relate to urbanization, where a formerly rural site becomes a site within the suburbs of a city.

Documentation of changes in atmospheric state variables and fluxes in response to LCC is extremely difficult and requires huge efforts. In the case of precipitation, for instance, accumulated precipitation can be used to visualize the impact. If a rural site, for instance, became an urban site, comparison with nearby other rural sites and urban sites can provide evidence for the land-cover impact on the atmospheric state variables and fluxes at the now urban site. Using rural–urban temperature differences to assess the impacts of urbanization on climate, however, may be inappropriate when the majority of stations are in or near cities, and only a few rural sites exit.

Detection of LCC impacts in routine measurements requires long time series in the area that underwent LCC, and the LCC must be documented well. Data of meteorological stations in the mountain–plain system of northeastern Colorado, for instance, evidences decreasing summer temperatures since the early 1980s that coincide with documented LCC (Chase et al. 1999). Zhou et al. (2004), for instance, documented evidences for the impacts of the rapid urbanization in southeast China since 1978 on climate. They found spatial pattern of near-surface temperature warming trends of 0.05 K per decade to be consistent with those of urban population changes and satellite-measured greenness (Fig. 3.3).

Due to the time constraints, the impacts of climate variability and the impossibility to experiment with a landscape, scientists tried to elaborate the impact of land-cover on the atmosphere by targeted field campaigns. These campaigns focused on the atmospheric response to one or more land-cover types. Several of these field campaigns were performed within the Global Energy and Water Cycle Experiment (GEWEX) framework.

Among the first campaigns to examine land-surface-atmosphere interactions was the Metropolitan Meteorological Experiment (METROMEX; Changnon 1980) in the 1970s. METROMEX studied the modification of mesoscale and convective rainfall by cities. METROMEX results showed that during summer, cities may increase precipitation by 5–25% compared to background values within the city area closest to the downwind and up to 50–75 km downwind of them. METROMEX results

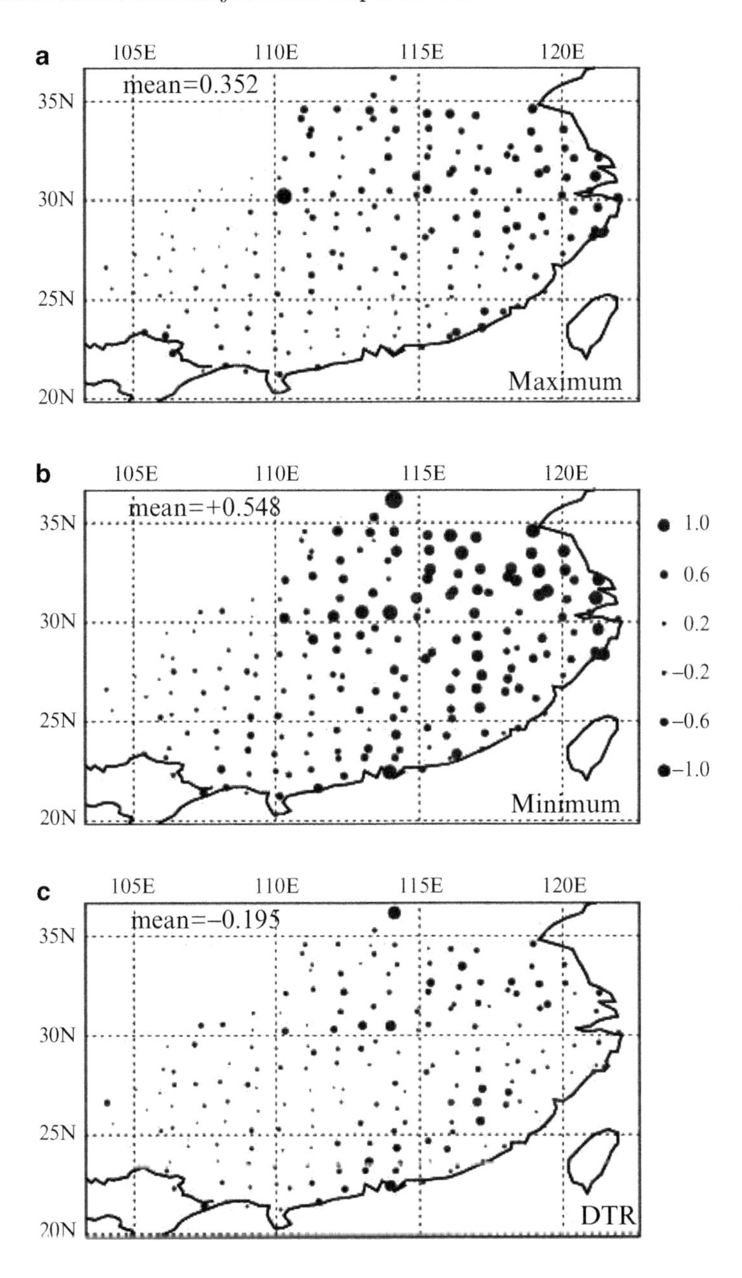

Figure 3.3. Observed winter temperature trends (in K per decade) over southeast China from 1979 to 1998 for maximum (**a**), minimum (**b**), and mean (**c**) (Modified after Zhou et al. (2004)

also indicated a relation between the areal extent of the city and the magnitude of urban and downwind precipitation anomalies.

The basic methodology of the GEWEX campaigns was to take measurements at a range of length scales (leaf scale, through the plot scale up to the mesoscale γ). Ground measurements were aggregated by means of atmospheric boundary layer (ABL) and aircraft measurements, remote sensing data and numerical modeling. Out of the large achievements of GEWEX, here only those relevant for land-cover impacts on the atmosphere are summarized briefly. GEWEX encompassed several experiments worldwide and hence covered a wide range of land-cover types and climate conditions.

The Hydrological and Atmospheric Pilot EXperiment and MOdélisation du BILan HYdrique (HAPEX-MOBILHY; e.g., André et al. 1990) was a GEWEX campaign carried out in France in 1986. It documented that the fluxes over various vegetation types have clearly distinguishable diurnal cycles. Forested areas that are very organized and extend more than 10 km in horizontal direction can develop an ensuing response in the ABL and may even initiate nonclassical mesoscale circulations (NCMC) in form of a "forest breeze."

The Amazon Region Micrometeorology Experiment (ARME; e.g., Shuttleworth 1988; Wright et al. 1992) provided measurements of the comparative near-surface climatology for large clearings and adjacent areas of undisturbed natural forest. Meteorological tower measurements showed that the atmospheric conditions in the lower forest space are decoupled from the atmosphere above the canopy, that is conditions in the canopy differ from those aloft. Interception of rainfall by the rainforest canopy is about 10–15%. About 50% of the incoming precipitation reevaporates, and about 25% stems from direct evaporation of intercepted rainfall. Precipitation-throughfall amount varies strongly in space. The measurements suggested that keeping the interception-collection gauges at the same place for a long time might result in significant pseudo-statistical errors, and hence should be avoided.

The First International Satellite Land Surface Climatology Project (ISLSCP) Field Experiment (FIFE; e.g., Sellers et al. 1992) took place in Kansas, USA, in 1987 and 1989. The area of this GEWEX-field campaign was covered mainly by prairie. During FIFE, moisture conditions varied over a wide range and allowed to observe the temporal evolution and spatial differences in dry-down of soils after heavy rains. Major findings were that stomata control the latent heat fluxes, and that seasonal variations in canopy aerodynamic temperature correlate linearly with canopy radiometric brightness temperature. Latent heat-flux measurements showed that the midday evaporative fraction and daytime average latent heat flux correlate strongly.

FIFE showed that land management affects the energy budget. During the growing season, shortwave reflectance measured over grazed sites was about 8% higher than over burned, non-grazed sites. It was 4–27% higher over mowed sites than over non-mowed sites. Near-infrared (NIR) reflectance was about 15% lower over grazed sites than burned sites and 18% lower over mowed sites than over non-mowed sites. Since soil albedo in the visible (VIS) spectral range exceeds the soil albedo in the NIR, any defoliation severe enough to expose the soil increases the reflectance in the VIS and decreases that in the NIR range.

In 1988, the LOngitudinal land-surface Transverse Experiment (LOTREX; Schädler et al. 1990) was carried out along one latitude (10°E) in Hildesheimer Börde, Germany, an area dominated by agricultural crops. Major findings were that the diurnal variations of latent and sensible heat flux and temperature differ among the various crops (sugar beet, wheat) and soil types. The experiment also gave hints on the formation and existence of internal boundary layers over boundaries of fields covered by different crops.

HAPEX-Sahel (Gash et al. 1997) took place in the Sahel in 1989. Major findings were that millet transfers a smaller proportion of the available energy to latent heat fluxes than tiger bush or fallow savannah. The seasonal accumulated precipitation is highest over fallow followed by millet and tiger bush sites. Transpiration of short grass decreased progressively once plant-available soil water was reduced by 30–40% due to the stomatal response to the water shortage.

HAPEX-Sahel suggested that data must be interpreted on a seasonal scale rather than a daily scale to explore land-cover impacts on the atmosphere. During the wet season when enough plant-available water exists in the root zone, the Sahel acts as a sink for atmospheric carbon. Under non-limiting soil-water conditions, the measured carbon dioxide (CO_2) exchange rates were lower than those typically measured over crops or grass-prairie. The erected leaves of Sahel grass intercept very little photosynthetically active radiation (PAR), and the structure and low LAI of the Sahel grass limit the assimilation rate. CO_2 exchanges decrease as plants experience water stress. Nocturnal emission of CO_2 from soils and plant respiration leads to maximum CO_2 concentrations of 360–390 ppm before sunrise. After sunrise, pronounced assimilation and vertical mixing reduce the atmospheric CO_2 concentrations to 320–330 ppm. As soil-water content decreases, the surface absorbs less CO_2 and the atmospheric CO_2 concentration increases (Monteny et al. 1997).

The SEBEX (Sahel Energy Balance Experiment; e.g., Wallace et al. 1991) took place in open forest and savannah of the Sahel in 1988–1990. SEBEX focused on direct measurements of available energy, evaporation, and sensible heat flux from contrasting Sahel land types to assess how LCC may affect these quantities. Major findings were that the relative contribution of the vegetation and soil to evapotranspiration correlates strongly with the temporal precipitation distribution and the growth of the savannah grasses. The growth is relatively insensitive to the precipitation amount. SEBEX confirmed the findings of HAPEX-Sahel that transpiration of short grass decreases progressively once plant-available soil water is depleted.

During the California Ozone Deposition Experiment (CODE, e.g., Pederson et al. 1995) in 1991, eddy-covariance measurements were performed to determine the vertical fluxes of ozone (O_3), CO_2, water vapor, sensible heat and momentum over grape vine, cotton, and senescent annual grassland. Additionally, aircraft measurements of fluxes were performed over these and other land-cover types at about 30 m above ground level (AGL). The latent heat fluxes at the cotton and vineyard sites exceeded those of the grassland site by a factor of 20 at noon. The sensible heat fluxes at the grass site were five and three times higher than at the cotton and vine site, respectively, which both had low Bowen ratios (i.e., the ratio of sensible to latent heat flux). At the grass site, ozone concentrations were higher and deposition velocities were lower than at the irrigated sites. Over the irrigated sites, CO_2, O_3, and water-vapor concentrations decreased when solar insolation decreased. The fluxes of CO_2, O_3, sensible and latent heat vary with land-cover heterogeneity. Ozone fluxes correlated modestly with the surface heterogeneity. Deposition of O_3 onto bare soil and nitrite oxide (NO) release from the surface affected the O_3-concentrations. Another major finding was that NCMC caused by extreme land-cover differences can affect the deposition of trace gases (Pederson et al. 1995).

The 1991 European Field Experiment in a Desertification-threatened Area (EFEDA; Bolle 1995) focused on such an area in Spain. In this semiarid region, the diurnal cycles of the surface-energy budget of nonirrigated summer crops, natural vegetation, and bare soil vary marginally from day-to-day and among each other. The water availability rather than the available energy or vertical water-vapor transfer limit evapotranspiration. During irrigation, the overall available energy (through the net radiation) increases due to the reduced albedo and surface temperature in response to the irrigation. The surface-energy budget of irrigated fields showed low Bowen ratios at the end of the irrigation. The day after the irrigation, the latent and sensible heat fluxes remained close to those during irrigation. The differences in latent and sensible heat fluxes due to the irrigation relaxed gradually into a drier mode in the following days.

The BOReal Ecosystem-Atmosphere Study (BOREAS; Sellers et al. 1997) took place in Saskatchewan and Manitoba, Canada, in 1994 and 1996. Major findings were that boreal forests are strong sources of sensible heat fluxes, but weak sources of latent heat fluxes as compared to midlatitude grasslands. During winter, boreal forests intercept efficiently shortwave radiation and convert it into sensible heat and outgoing long-wave radiation. In spring, the late thawing of the soil keeps the root system frozen and cuts transpiration off leading to the very high sensible heat fluxes over boreal forests. In wetlands, the forest canopy intercepts almost all of the available energy. Consequently, the underlying wet soils or mosses hardly affect the surface energy balance at all. This partitioning of incoming energy generates a dry and warm lower troposphere with a deep turbulent ABL over the boreal forest during the growing season. This finding was quite unexpected for a seemingly moist biome (Sellers et al. 1997).

Albedo has a distinct annual cycle that affects the energy budget. Without snow-cover, summer daily average albedo was about 0.2, 0.15, and 0.083 over grass, aspen forest, and coniferous forest, respectively. Winter daily averages of albedo over snow-covered grass, aspen, and conifer areas were 0.75, 0.21, and 0.13, respectively. Smoke and associated aerosols reduced frequently the incoming solar radiation during the growing season, and hence affect the surface-energy budget (Sellers et al. 1997).

The BOREAS trace-gas studies indicated that dry soils with fast nitrogen cycling covered by aspen or jack pine could act as strong methane (CH_4) sinks, while wet soils covered with fern or ponds are more likely to be methane sources.

In 1997 and 1999, the Cooperative Atmosphere Surface Exchange Study (CASES; LeMone et al. 2000) took place in Oklahoma. CASES revealed that soil moisture and ground heat fluxes lead to differences across different vegetative regimes. Clear-day data of CASES showed that latent heat fluxes increased over grassland as grass green-up occurs and the photosynthetic activity increases gradually. Cloud-cover can modify the surface-energy budget strongly and the various land-cover types respond differently to variation in cloudiness. Cloud-cover enhanced the variability of latent heat fluxes at sites with growing winter wheat stronger than at sites with pasture or grassland. However, cloud-cover affected hardly the sensible heat fluxes over either of these sites.

Various observational studies evidenced the formation of NCMC in regions of huge land-cover contrasts (Segal et al. 1988). For instance, during BOREAS under light synoptic scale wind conditions, snow breezes were observed across a lake of 4 km in extent. A temperature contrast between the frozen, snow-covered lake and its adjacent forest built after sunrise. The resulting pressure gradient established onshore wind that persisted into the afternoon (Taylor et al. 2007). During CODE, vegetation breezes established between irrigated fields and very dry (senescent) annual grassland in the San Joaquin Valley (Pederson et al. 1995). For the Sahel, Taylor et al. (2007) reported low-level divergence over wet soil and convergence in relatively drier areas within a wet zone extending 160 km after rainfall events. Statistical analysis showed coherence of soil moisture and atmospheric wind patterns on wavelengths down to 20 km. According to well-resolved aircraft observations, ABL temperature, moisture, and height were up to 3 K lower, 3 g/kg higher, and 50% lower over wet soil than over adjacent relatively drier areas.

Table 3.1. Surface characteristics of various land-cover types.

Land-cover	ε	α	z_0 (m)	LAI	$r_{st,min}$ $(m^{-1}\,s)$	z_{root} (m)
Bare, sparsely vegetated land	0.84–0.91	0.15–0.3	0.005–0.01	0–1	999[c]	0.5
Coniferous forest	0.97–0.98	0.1–0.15	1–4	3.2–9.2	43–175	3.7–3.9
Cropland	0.95–0.97	0.18–0.25	0.05–0.17	0.2–8.7	25–110	0.32–2.35
Deciduous forest	0.95–0.97	0.13–0.2	0.8–3.5	3.6–8	43–232	2.9–3.9
Grassland	0.9–0.95	0.16–0.3	0.03–0.08	0.04–5	25–77	0.32–2.6
Heather, bushland	0.95–0.97	0.15–0.2	0.3–0.4	0.4–4.6	42–330	0.5–4.8
Meadows, wetland	0.98–0.99	0.14–0.25	0.03–0.3	2.5–8.4	25–77	0.5–1.81
Rainforest	0.97–0.98	0.12–0.15	0.55–2	0.6–9.3	100–120	2.9–3.9
Steppe, savannah	0.97–0.99	0.19–0.3	0.1–0.2	1–5.1	120–167	15.0
Shrubland	0.87–0.97	0.1–0.2	0.03–0.55	0.4–5.1	200–330	4.8–7.0
Tundra	0.94–0.99	0.15–0.25	0.03–0.06	0.1–2.2	100–150	0.5–1.81
Wooded tundra	0.93–0.97	0.16–0.18	0.05–0.06	3.1–5.3	150–200	1.81
Urban	0.85–0.95	0.1–0.27	0.8–2.5	–	–	2.9
Water bodies	0.972–0.993	< 0.05[a]	$f(v)$[b]	–	–	–

The symbols ε, α, z_0, LAI, $r_{\mathrm{st,min}}$, and z_{root} represent emissivity of long-wave radiation, albedo in the visible spectral range, roughness length, leaf-area index, minimum stomatal resistance, and root length. Values indicate the range of values found in the literature cited in this book

[a]Except in sun glint

[b]Function of wind-speed and wave height

[c]or any other number that serves in the model as upper value for high (close to have "infinity") resistance

All these experiments also increased the knowledge on the variability of soil, and surface vegetation parameters. Table 3.1 summarizes typical ranges of vegetation parameters as obtained from various field campaigns and used in numerical modeling.

3.3 Mechanisms

The various field experiments elucidated the importance of land-cover for the modulation of the near-surface air conditions. For LCC to affect the atmosphere, the large-scale (synoptic scale) basic flow has to be favorable. Obviously, various LCC influence the atmosphere most for synoptic scale flows between 5 and 10 m/s (Mölders 1999a).

The field experiments also triggered the development of sophisticated land-surface and ABL models for help in interpretation of the measurements. The related model development improved NWP, air-quality, and climate models.

A major shortcoming of field experiments is that they cannot be repeated under the exactly same conditions. Consequently, scientists have to rely on similar soil and large-scale atmospheric conditions for the various sites of different land-cover and have to assume that the differences in measured fluxes are only due to the different land-cover. Numerical modeling methods, however, permit experimenting with the atmosphere. In a model, processes can be switched on or off. Parameters like land-cover can be exchanged easily, while using otherwise the same initial and boundary conditions. The difference between the results from simulations without and with the LCC at steady state then permits an assessment of the impact of the LCC, while the uncertainties generated by other factors are to be negligible.

For these reasons, numerical modeling is an important tool to elaborate the mechanisms that cause the responses to LCC and to examine land surface-atmosphere interactions. Of course, models deployed for LCC and process studies must include sophisticated, well-evaluated land-surface, cloud and ABL models.

Experiments and modeling studies showed that on days with weak synoptic forcing ($v < 5\,\mathrm{m/s}$) mesoscale circulations can develop. Such circulations are forced mainly by variability in orography, surface type, land-cover, soil temperature, and moisture and soil type. At the interface earth-atmosphere, biogeophysical processes regulate and hence surface characteristics govern the fluxes of momentum, water, trace gases, and heat. The models confirmed the experimental indications that the partitioning of incoming energy between sensible and latent heat differs

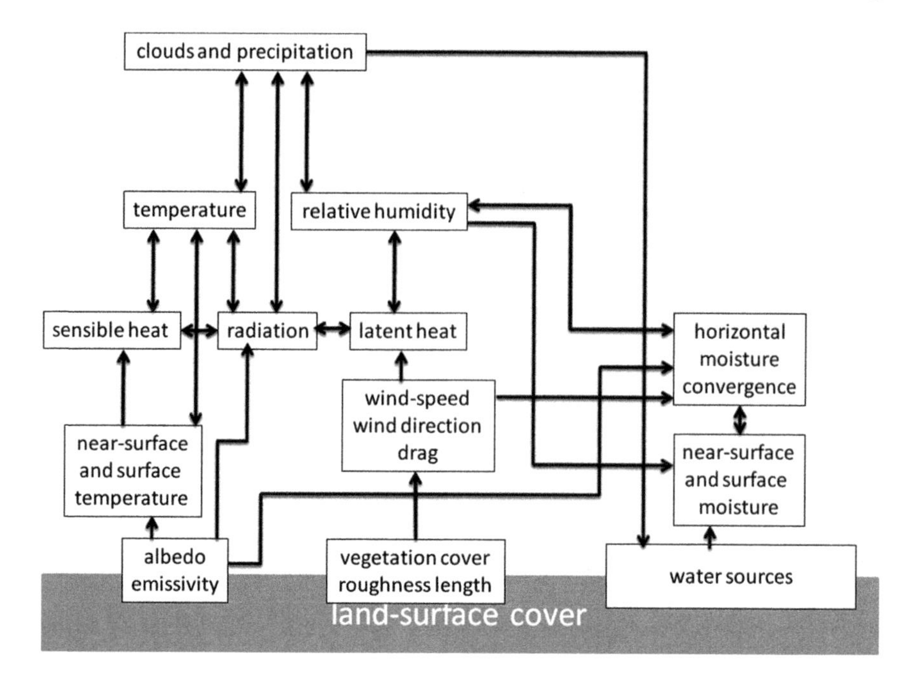

Figure 3.4. Schematic view of biogeophysical feedback processes in response to LCC

appreciably among land-cover types. The fluxes modify the state variables that again depend on the fluxes and modify them. Thus, LCC may affect the microclimate by altering the surface energy budget, the dynamics, including vertical mixing and ABL structure, the thermodynamic environment for convection and moist static energy, and runoff by nonlinear feedback mechanisms.

3.3.1 Changes of the Surface Fluxes

As land-cover changes, surface properties like heat capacity, heat conductivity, albedo, roughness length, LAI, maximum evaporative conductivity, heterogeneity, etc. change (cf. Table 3.1). Altered heat capacity and/or water availability modify the partitioning of energy into sensible and latent heat fluxes characterized by the Bowen ratio, while a modified albedo alters the fraction of solar radiation reflected back to space (Fig. 3.4). Radiative responses to LCC are the altered Bowen ratios, surface and near-surface temperatures. According to Charney et al.'s (1975) hypothesis, an increase (decrease) in albedo decreases (increases) net radiation and temperature. An increase (decrease) in albedo decreases (increases) the surface moist static energy flux. However, as will be discussed later exceptions exist.

Some surface parameters influence the energy fluxes stronger than others do. Collins and Avissar (1994) investigated the relative contribution of individual land-surface parameters to the variance of energy fluxes by Fourier amplitude sensitivity tests. They found that the distribution of relative stomatal conductance and surface roughness explain most of the variability of surface-heat fluxes. Many modeling studies suggested that the altered albedo and roughness length due to LCC determine the magnitude of the atmospheric response strongly.

Changes in roughness length due to LCC cause dynamic responses (Fig. 3.4). Increased (decreased) surface roughness may increase (decrease) significantly the vertical mixing and affect the structure of the ABL. Increasing of the roughness length may also shift the precipitation distribution upstream. Various observational studies showed that fronts reduce their speed of propagation over large conurbation as compared to their speed of propagation over the rural area (e.g., Loose and Bornstein 1977).

Land-cover changes may alter the uptake of particles and the emission of trace gases as well as the deposition of particles and trace gases. The altered atmospheric composition may further affect aerosol formation by gas-to-particle conversion. The altered aerosol spectrum and concentration of radiative active gases modify the surface-energy budget in particular and the radiation budget in general. Aerosols and gases scatter and absorb radiation in large parts of the spectrum and hence modify the surface-energy budget and the atmospheric temperature profile directly.

The changes in surface energy, moisture, and momentum fluxes due to altered surface properties are referred to as biogeophysical feedbacks or effects. The changes in surface-energy, moisture, trace gas and momentum fluxes, and the corresponding variables of state are the primary changes in response to LCC.

3.3.2 Secondary Changes in Response to Land-Cover Change

The primary responses (modified thermal stratification, evapotranspiration, etc.) may cause secondary changes (modified cloud processes, dynamics, insolation, etc.). The altered surface characteristics in response to LCC namely cause small differences in ABL temperature and moisture. Such variations, even if they are within the range of observational errors (1 K, 1 g/kg), can be decisive for no initiation or development of intense convection (Crook 1996).

The altered fluxes and state variables in response to LCC mean a change in atmospheric stability and buoyancy. Slight changes in temperature and/or water-vapor content due to LCC may shift cloud properties and affect the time of onset of cloudiness and precipitation. METROMEX, for instance, showed that clouds over the urban area had higher cloud bases and cloud tops than clouds in the adjacent rural area. The relatively warmer air over the city has to rise higher to reach the lifting condensation level (LCL) than the relatively colder rural air.

The physical concept of these changes can be described by the vertical and pseudo-adiabatic lifting of an air parcel of unit mass in a hydrostatically balanced environment. The air parcel accelerates according to

$$\frac{\mathrm{d}w}{\mathrm{d}t} = \frac{g}{T_\mathrm{v}} \left(T_\mathrm{v,p} - T_\mathrm{v}\right) \tag{3.1}$$

While the pressure acting on the parcel adjusts to that of the environment. Here, w, g, $T_\mathrm{v,p}$, and T_v are the vertical velocity of the air parcel, gravity acceleration, the virtual temperatures of the air parcel, and environment, respectively. Multiplication of Eq. 3.1 with the differential of the height and use of the hydrostatic equation yield

$$\begin{aligned} \mathrm{d}z\frac{\mathrm{d}w}{\mathrm{d}t} &= \frac{\mathrm{d}z}{\mathrm{d}t}\mathrm{d}w = w\,\mathrm{d}w = \frac{\mathrm{d}}{\mathrm{d}t}\frac{w^2}{2} = \frac{g}{T_\mathrm{v}}\left(T_\mathrm{v,p} - T_\mathrm{v}\right)\mathrm{d}z \\ &= -R_\mathrm{d}\left(T_\mathrm{v,p} - T_\mathrm{v}\right)\mathrm{d}\left(\ln p\right) \end{aligned} \tag{3.2}$$

where R_d is the gas constant for dry air and p is air pressure. Integration from the level from where parcel started rising, p_OL, to the level of free convection, p_LFC, provides the kinetic energy change per unit mass

$$\begin{aligned} w_\mathrm{LFC}^2 - w_\mathrm{OL}^2 &= -2\int_{p_\mathrm{OL}}^{p_\mathrm{LFC}} R_\mathrm{d}\left(T_\mathrm{v,p} - T_\mathrm{v}\right)\mathrm{d}\left(\ln p\right) \\ &= 2\int_{p_\mathrm{LFC}}^{p_\mathrm{OL}} R_\mathrm{d}\left(T_\mathrm{v,p} - T_\mathrm{v}\right)\mathrm{d}\left(\ln p\right) \end{aligned} \tag{3.3}$$

If between the original level and LCL the parcel's virtual temperature is lower than that of the ambient air, convective inhibition (CIN) will occur. This "negative energy" represents the stability barrier that the air parcel has to overcome to reach the level of free convection (LFC).

Figure 3.5. Aerial photograph of the area along rabbit-proof fence in Australia that separates the native vegetation and farmland. Clouds form randomly over the dark, relatively moist native vegetation (Modified after `http://www.nytimes.com/2007/08/14/science/earth/14fenc.html`)

For free convection to occur, w_{OL} must be at least greater or equal to $(-2\mathrm{CIN})^{0.5}$.

Under same synoptic conditions, the heating of an air parcels close to the surface depends on the surface characteristics, that is land-cover. The air parcel's temperature may be different after LCC. Thus, LCC may change the amount of energy per unit mass required for lifting the air parcel vertically and pseudo-adiabatically from its original level to its LFC. Consequently, LCC can affect whether or not free convection occurs and/or the timing of its onset (e.g., Fig. 3.5).

Aloft the LFC, the parcel's temperature exceeds that of the ambient air until the two temperature curves intersect again. This region of "positive energy" is known as the convective available potential energy (CAPE). CAPE represents the amount of buoyant energy that is available to accelerate the air parcel vertically and is sensitive to LCC.

The strong vertical motions within clouds redistribute trace gases and particles vertically. Clouds may lift trace gases and particles into the free troposphere where these species may become subject to long-range transport. Consequently, LCC that enhance or diminish buoyancy may not only affect air quality in the immediate vicinity downwind of their occurrence, but also on large-scale via long-range transport.

LCC that lead to a shift in the height of the 0°C-isotherm, affect air chemistry via altered gas-phase and aqueous-phase equilibria and the partitioning of the cloud-microphysical processes between the warm and cold paths of precipitation formation may differ appreciably due to the altered properties of the lifted air. Usually temperature decreases with height. Thus, for saturated air that due to LCC reaches higher levels this means an increased likelihood of ice-crystal formation. Since the saturation-vapor pressure over ice is lower than that over water, ice-crystals may form at the cost of evaporating cloud droplets (Bergeron-Findeisen process). This cold path of precipitation formation is much more efficient than the warm path of precipitation formation by collection and coalescence of droplets of different size. In response to LCC, the greatest changes in supercooled cloud-water and rainwater occur in the ABL at the flight altitude of small airplanes, and hence may pose a threat to private air traffic (Mölders and Kramm 2007).

The release of latent heat and consumption of heat during the various phase-transition processes modify the temperature profile. The modification of the paths of cloud microphysical processes and the level where they occur, further affect the vertical temperature profile. Due to the interaction between microphysics-dynamics, updrafts and downdrafts may occur at different locations and with different strength than in the unperturbed landscape. Consequently, the vertical mixing is changed further.

Clouds are typically ice nuclei (IN) limited. Altered gas-to-particle formation and/or emission of hygroscopic aerosols in response to LCC affect the number and spectra of cloud condensation nuclei (CCN) and/or IN. If LCC increase the number of IN, the ice-particle concentrations will increase and ice-particle sizes will decrease; the cloud may glacify over a relatively narrow altitude range.

LCC, which reduce the number of CCN/IN, shift the particle spectrum toward larger particles and accelerate the onset of precipitation. The opposite occurs for increased numbers of CCN/IN. If under saturated conditions CCN (IN) compete for excess water vapor, smaller, but more cloud-droplets (ice-crystals) will form. The reduced cloud-droplet (ice-crystal) sizes not only diminish coalescence (riming), thereby decreasing precipitation–formation efficiency and increasing cloud lifetime, but also enhance cloud albedo. The modified cloud albedo leads to radiative feedbacks and modification of the vertical temperature profile and again the surface-energy budget. Since cloud albedo affects the photolysis rates, any changes in cloud albedo modify the gas-phase chemistry below, within and above clouds.

Any change in the vertical profiles of temperature, CCN/IN, water vapor and soluble gases influence phase-transition processes with further consequences for the temperature, CCN/IN, and water-vapor profiles.

A change in hydroscopic aerosol concentrations in response to LCC affects the atmosphere via changes in phase-transition processes (thermodynamic effects). In addition, the reduced droplet size delays freezing. All these thermodynamic effects contribute to changes of the total thermal energy within the cloud and its environment.

Aerosols absorb solar radiation, which may lead to evaporation/sublimation of liquid/solid cloud particles (semi-direct aerosol effect). The increased aerosol amount and cloud optical thickness alter the surface-energy budget as they modify net surface solar radiation. On the local scale, aerosols residing in the troposphere reflect sunlight to space leading to daytime surface cooling; they emit infrared radiation downward to the surface (infrared effect). At night, this process leads to surface warming. The infrared effect can increase air temperatures and evaporation/sublimation of cloud particles. Any changes in the aerosol distribution caused by LCC will influence the semi-direct aerosol effects.

Clouds reduce the incoming radiation. The area affected by the cloud depends on cloud type. Stratiform clouds, for instance, affect areas of several $100 \times 100\,\mathrm{km}^2$, while convective clouds only affect a small area in a certain angle underneath. In this case, the shadow intensity, extent, and location depend on cloud-base height, extension, and solar zenith angle. In the cloud shadow, evapotranspiration is reduced. As evapotranspiration is related to the latent heat fluxes via the latent heat of evaporation, clouds affect the surface-energy budget and how the available radiation energy at the surface is partitioned between sensible and latent heat. This means that if LCC lead to enhanced cloud formation as a secondary impact, the clouds may reduce the evapotranspiration via the cloud-evaporation feedback. However, this feedback process does not necessarily offset the changes caused by the LCC.

The secondary changes cause further differences in cloud distributions. Consequently, the perturbations due to LCC can propagate appreciable distances form the area of LCC. Secondary changes in response to LCC may become large enough to affect the mean large-scale flow and mesoscale circulation pattern (e.g., D'Almeida et al. 2007).

3.3.3 Teleconnection

Land-cover changes may occur at various extensions. Most LCC are of local scale, but may affect weather and climate far in the downwind by secondary processes. The early studies on LCC mainly focused on the atmospheric impacts within the region of LCC and/or their downwind. In the 1990s, some studies introduced the idea that extended LCC may affect global climate through teleconnections (Chase et al. 2000) where teleconnection refers to climate anomalies that are associated with each other, but occur in different regions. Consequently, climatic changes in response to extended persistent LCC will not be restricted to the region of LCC and its immediate downwind, but teleconnect to regions far remote from the LCC where the teleconnection becomes the direct forcing. Large-scale tropical LCC and the associated circulation changes, for instance, were found to cause persistent atmospheric changes similar to those of sea-surface temperature (SST) anomalies associated with El Niño/Southern Oscillation (ENSO) that may cause anomalies at midlatitudes (Werth and Avissar 2002).

Studies on Amazonian deforestation performed with General Circulation models (GCMs) generally show atmospheric responses in areas adjacent to and in the immediate downwind of the LCC and reduction of precipitation on global average. Werth and Avissar (2002) examined with the Goddard Institute for Space Studies global climate model how far away the atmospheric responses to Amazonian deforestation may reach. In midlatitudes, the atmosphere still responds statistically significantly to the deforestation, but weaker than in Amazonia. In some midlatitude regions, precipitation decreases throughout most months and usually the largest decrease occurs during the rainy season (Fig. 3.6). In the Midwest Triangle and Dakotas region, for instance, Amazonian deforestation dampens the summertime peaks of the annual precipitation, especially in July; evaporation and cloudiness decrease year-round and most strongly in summer. The authors concluded that the atmospheric response over Amazonia modifies the Hadley circulation regionally and hence the moisture advection from the Tropics. The decreased latent heat release due to deforestation reduces temperature in the upper Westerlies. The spatial shift in and reduction of convection together with the altered compensating circulations modify the magnitude and pattern of high-level tropical divergence that drives the zonal jet in midlatitudes (Chase et al. 2000). The atmospheric response over Amazonia acts as a source for low-frequency tropical waves. These waves propagate to the Extratropics and establish a teleconnection between the Amazon region and the regions remote to the LCC.

Teleconnections also occur for mid- or high-latitude LCC. Li (2007) examined the remote impacts of LCC in the Yukon, Ob, St. Lawrence, and Colorado region and their interaction on evapotranspiration and precipitation. The assumed LCC were local changes of forest to grassland, and grassland to cropland in the Yukon and Ob regions. In the Colorado region, local changes of grassland to shrubland or cropland, and forest to grassland were assumed. In the St. Lawrence region, she

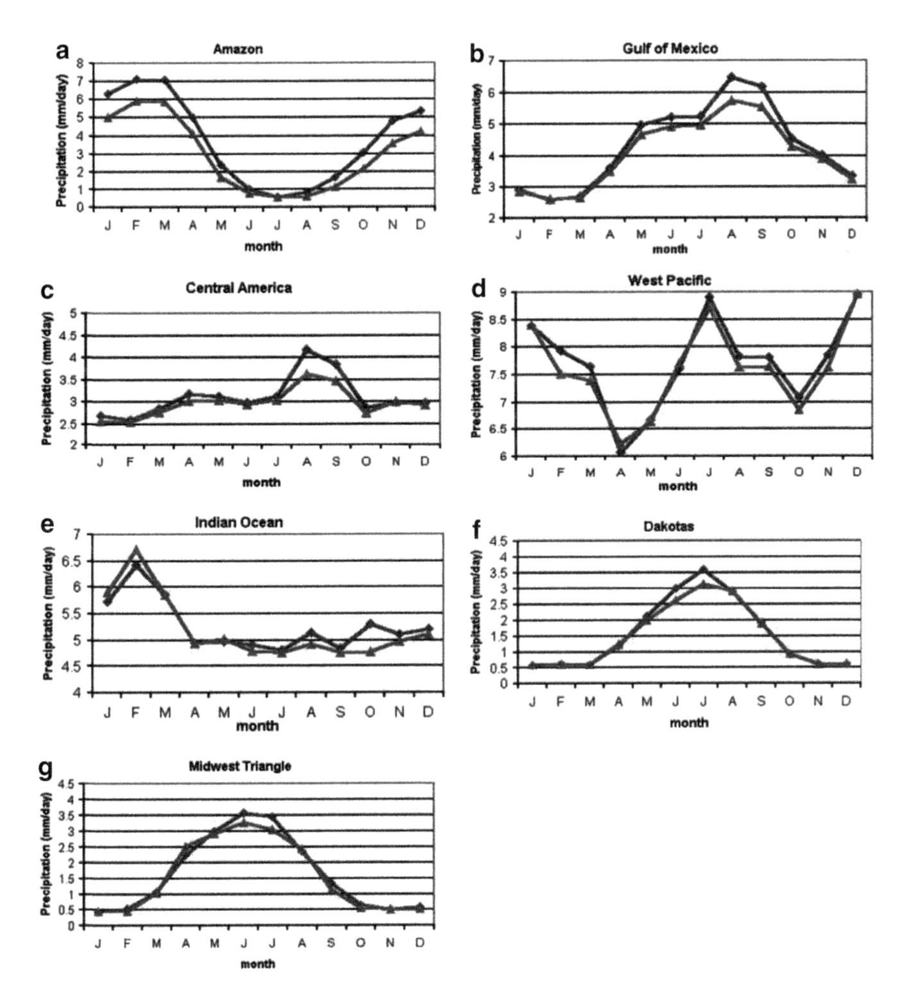

Figure 3.6. Impact of teleconnections in response to deforestation of the Amazon rainforest as seen in the annual precipitation cycle in the (**a**) Amazon, (**b**) Gulf of Mexico, (**c**) Central America, (**d**) West Pacific, (**e**) Indian Ocean, (**f**) Dakotas, and (**g**) Midwest Triangle. Lines with diamonds and triangles indicate the precipitation as obtained for the ensemble mean 8-year average for the reference case with vegetation at its 1950 levels and the case assuming deforestation of the Amazon rainforest at the benefit of a mixture of shrubs and grassland (Modified after Werth and Avissar 2002)

assumed land abandonment and deforestation at the benefit of grassland and cropland. Her suit of simulations with the fully coupled Community Climate System Model assumed LCC in only one of these regions as well as concurrently in all four regions. In most months, the atmospheric responses to the LCC in the four regions interacted via teleconnection.

This interaction enhanced the overall atmospheric response as compared to the cases with LCC in the Ob or St. Lawrence region alone. The opposite was found for the Yukon and Colorado regions. The interactions of the responses to LCC alter the magnitude of regional change rather than the spatial pattern or extension of the response.

Heat and moisture advection as well as secondary effects propagate the impact of LCC to remote areas (Li 2007). The strength and/or time of LCC-induced teleconnections varies among the regions of LCC. Precipitation and evapotranspitation changes correlate strongly in summer when the land-surface conditions govern convective precipitation. In summer, the long residence time (i.e., the ratio of precipitable water to precipitation) of water vapor permits the penetration of precipitation and evapotranspiration changes in the downwind of the LCC. The changes in downwind precipitation trigger further changes in their downwind via modified water availability for evapotranspitation and moisture advection. Generally, the remote changes in precipitation exceed those in evapotranspiration (Li 2007).

How far teleconnections establish depends on the region of LCC, the remote regional climates, season, and favorable large-scale conditions. Favorable large-scale conditions require strong atmospheric transport and residence times of about 10 days. In Li's (2007) experiments, for instance, the atmospheric responses reach the farthest for the LCC of the Ob region during winter, for the St. Lawrence region during summer, and for the Yukon and Colorado regions during spring and fall. In the St. Lawrence region, the large-scale meteorological conditions control strongly evapotranspiration from September to March. Thus, the LCC affect hardly the spatial distribution of evapotranspiration. In response to the LCC, the high-pressure system over the south to the northeast of the Colorado region retreats from February to April. From September to November, the LCC cause an opposite location change of the high-pressure system (Li 2007).

3.4 Scales

The slight changes in atmospheric state variables and fluxes due to LCC can be interpreted as perturbations of the large-scale flow. We can consider the atmospheric state variables or fluxes χ as the sum of the quantity representing the mean undisturbed large-scale condition $\bar{\chi}$, the mesoscale perturbation $\hat{\chi}$, and the perturbation due to the LCC χ'.

Any persistent LCC affects local (i.e., in the area of the actual LCC) and regional (areas adjacent to the LCC) climate on the short and long term. Consequently, investigations looked at the impacts of LCC at various temporal and spatial scales as well as at possible nonlinear feedbacks. Many land-surface properties affect the atmosphere differently at different time in the diurnal or annual cycle. Mesoscale simulations wherein one land-surface parameter at a time was changed within its natural range of variability for all grid-cells showed that except for wind and cloud and precipitation particles, land-surface parameters influence the atmosphere least at night and in the early afternoon than at other times (Mölders 2001). A change from agriculturally used land to grassland, for instance, hardly affects the atmosphere in winter in snow-covered regions, while the fluxes of momentum, trace gases and sensible and latent heat over grassland and agriculturally used land differ strongly during the growing season. Even for agriculturally used land, the crop-type grown may cause huge differences in the temporal evolution of the surface-energy budget components. Not only albedo, emissivity, LAI, and height of crops (roughness length) change during the growing season but also demands of water amount and the time of water demand differ among species. Corn, for instance, has its highest water demands at a time when wheat is harvested already. Consequently, evapotranspiration from a cornfield may be much higher for same meteorological and soil conditions than that of the harvested wheat field.

For LCC to influence weather and climate, the large-scale conditions have to be favorable. This means that LCC impacts on the atmosphere differ under different atmospheric conditions. Often LCC impacts are hard to identify in just one quantity (e.g., precipitation) without modeling due to secondary responses, spatial offsets of responses, and/or differences in the large-scale forcing. Lauwaet et al. (2010), for instance, examined the influence of LCC on the evolution of a mesoscale convective system (MCS) in West Africa, but found no clear response of precipitation. They performed simulations with a high-resolution regional atmospheric model wherein vegetation fraction was increased/decreased by 10%, 20%, and 30% between 10°N and 15°N. In the days before the MCS occurs, the vegetation changes influence CAPE and surface fluxes clearly. They found that the cold pool dynamics of the MCS alter the moisture conditions in the ABL and hinder a clear signal of the assumed vegetation changes on precipitation.

Possible atmospheric responses depend on the region that experiences LCC. Regions dominated by advection of precipitation (e.g., by frontal passages) are less sensitive to LCC than regions dominated by local recycling of previous precipitation. The availability of water is the single most important prerequisite for land-plant productivity. Thus, LCC that modify the water supply to the atmosphere thereby alter appreciably the local recycling of precipitation, and cloud and precipitation formation.

3.4.1 Inhomogeneity

Atmospheric models require vegetation parameters (e.g., albedo, minimum stomatal resistance, LAI, roughness-length, emissivity) to describe the land-cover. The various parameters are empirical and stem from various measurements for the same vegetation type, that is they were not determined within the same experiment. Land-surface parameters may vary over several orders of magnitude, even within a small plot of same vegetation, that is they are not natural constants. Even when taken within a 3-m radius of a seemingly homogeneous potato stand under relative constant atmospheric conditions, data of over 1,000 measurements showed such variability for relative stomatal resistance and its frequency (Avissar 1993). The high variability results from the locally different and continuously changing micrometeorological conditions to which stomata assimilate rapidly. Within the same canopy, micrometeorological conditions differ due to leaf inclination, orientation, shading, ventilation, and location within the stand as well as stand height and density (e.g., Avissar 1993).

The uncertainty in empirical vegetation parameters affects the simulated fluxes determined therewith. Any model results (e.g., fluxes, variables of state, cloud- and precipitation-formation) based on fixed land-surface parameters inherit uncertainty due to the uncertainty of these parameters via error propagation. Various studies showed that the arithmetic averages of the surface characteristics fail to provide the area-aggregated flux (e.g., Fig. 2.6). Instead, the area-mean latent heat flux has to be calculated for the various stomatal resistances. The resulting fluxes have to be frequency-weighted to determine the area-representative flux. Using an average stomatal resistance instead of the described procedure can lead to 10–20% error in latent heat fluxes (Mölders 2001).

Mölders (2005) used Gaussian error-propagation principles to assess the impact of uncertainty of land-cover parameters on simulated surface fluxes. This method bases on the idea that simulated fluxes will have a stochastic error (or standard deviation) due to the parameters' standard deviations. Applying this method and typical uncertainty ranges for the parameters yields relative errors of net radiation, sensible, latent, and ground heat flux of, on average, 7%, 10%, 6%, and 26%, respectively. Parameters affecting net radiation critically are vegetation fraction and emissivity. Sensible and latent heat fluxes are very sensitive to vegetation-fraction uncertainty. The stomatal resistance for most land-cover types is most sensitive to minimum stomatal resistance and LAI. The Gaussian error-propagation principles revealed that the empirical parameters with the highest relative error do not necessarily determine the standard deviation of the predicted fluxes.

3.4.2 Heterogeneity at Various Scales

Planning and management issues related to LCC motivated most local-scale studies on the atmospheric impacts of LCC. Most of these studies involved just climate data. Planning studies that involved mesoscale modeling suggest that under favorable large-scale conditions, local scale simple LCC may produce dynamic and kinetic perturbations on the conditions over and downwind of the LCC even on the short-term (up to 24 h). If the atmosphere is saturated, cloud and precipitating particles can differ significantly over and downwind of the LCC.

Mesoscale LCC studies showed that regional LCC not only have a local response but may also influence the weather and climate far beyond the region where they actually occur due to secondary changes. These studies also showed that within the region of LCC it takes some time for the regional climate to adjust to the LCC.

For large-scale LCC, the temporal scale gains importance and may differ for the various aspects of the land-atmosphere system. In the case of deforestation, for instance, surface climate needs between 1 and 2 years to establish fully to the new conditions of the deforested area. In the root zone, soil moisture still decreases in year 3. Soil moisture variability and total cloud amount continue to increase in the 6-year simulation (Henderson-Sellers et al. 1993).

Due to these facts, different types of models have been deployed depending on what kind of aspects were to be investigated. Extreme LCC (e.g., deforestation of an entire region) and their impacts on large-scale (e.g., monsoon, Walker- and Hadley-circulation, Rossby waves) or classical mesoscale circulation or NCMC pattern, including impacts on precipitation, are typically examined with GCMs. Mesoscale meteorological processes related to organized land-cover occur at the non-resolvable (subgrid) scale in GCMs. Note that to resolve the atmospheric response to land-cover differences in models, the land-cover must be the same in at least two adjacent grid-cells. In nature, land-cover variations being of subgrid-scale with respect to the resolution of the model modify the state variables and fluxes represented by a grid-cell or grid-area.

Thus, regional LCC may induce thermally induced NCMC that are of subgrid-scale with respect to the resolution in GCMs. The synoptic scale wind-field may interact with such land cover–induced circulations. Such NCMC may become as intense as a sea-breeze circulation when extended areas of unstressed dense vegetation are adjacent to bare land (Segal et al. 1988). Thus, they may influence significantly the onset of cloud formation and may modify potentially precipitation intensity and distribution (Chen and Avissar 1994). The inclusion of subgrid-scale variation of land-cover often leads to less severe consequences of LCC than suggested by GCMs (Shuttleworth 1988). Mesoscale models allow the assessment of LCC and land–atmosphere feedbacks at scales unresolved by GCMs as they represent atmospheric instabilities induced between areas of different land-cover better than GCMs. The impacts of such instabilities differ strongly from the results obtained by GCMs.

3.4.3 Impact of Heterogeneity

Most LCC alter the degree of heterogeneity of the landscape. Depending on the type of LCC, the number of ecosystems and the extent of areas with same land-cover may change. The degree of heterogeneity can be assessed by

$$\delta_{\mathrm{het}} = 0.5\left(\frac{\Phi}{\Phi_{\mathrm{max}}} + \frac{X-1}{X_{\mathrm{max}}-1}\right) \tag{3.4}$$

where Φ is the length of boundaries between areas of different land-cover, and X is the number of ecosystems occurring in the landscape. Furthermore, Φ_{max} and X_{max} are the total lengths of the possible boundaries and maximum number of land-cover types that, in principle, may exist in the landscape in the respective climate region or in the dataset considered. Following this definition, heterogeneity is maximal for $\delta_{\mathrm{het}} = 1$. For gridded data (e.g., GIS, USGS land-cover data, SiB land-cover data), this would mean that each grid-point is the boundary to another land-cover type on all sides. Complete homogeneity ($\delta_{\mathrm{het}} = 0$) exists if only one land-cover type occurs; hence, no boundary exists in the landscape of interest.

Modeling studies assuming simple LCC suggest that the magnitude of the atmospheric response to LCC is independent from the fraction of the region that experiences LCC (e.g., Friedrich and Mölders 2000). Instead, the magnitude of response depends on the size of patches with LCC, the size of patches with homogeneous land-cover after the LCC, the heterogeneity of the landscape, and the contrast in the hydrologic and thermal behavior of the new and old land-cover.

The atmospheric and surface responses to a conversion from moderately or strongly evapotranspirating surfaces to weakly evapo(transpi)-rating surfaces, for instance, depend on the size of the changed patches. Patches of same land-cover must exceed a certain size for distinct patterns of upward or downward motions to establish. Consequently, in a landscape, a given fractional LCC will influence the local conditions more strongly if the resulting patches of same land-cover have a large than small extent. This means the size of the patch determines strongly the magnitude of the atmospheric response.

The horizontal distribution of sensible heating is important for convection development. Under favorable large-scale flow conditions, the patch with a given land-cover type must exceed several kilometers to produce a discernable atmospheric response (Shuttleworth 1988). Organized heterogeneity – like planting bands of vegetation with width of 50–100 km in semiarid regions – can increase convection and precipitation. The increases exceed the increases associated with uniformly planting the area (Anthes 1984).

Homogeneous landscapes produce less precipitation than landscapes with relatively large discontinuity. The latter induce a negative feedback to reduce the discontinuity. In homogenous landscapes of only one land-cover type without distinct differences in soil characteristic and topography, convection occurs randomly distributed. If the area experiencing LCC to just one type becomes large enough ($> 600\,km^2$ or so), mesoscale circulations may be modified or NCMC may be initiated due to thermal differences among adjacent extended surfaces (e.g., Changnon 1980; Segal et al. 1988; Mölders and Kramm 2007).

3.5 Land-Cover Change Impacts

The following sections provide a discussion of the impacts of various types, kinds, and sizes of LCC on the physical and chemical states of atmosphere. This discussion includes greenhouse gas (GHG) release from LCC that is commonly known as "carbon debt." Despite differences with respect to the scales addressed, all studies indicate substantial or even significant differences in the variables of state and fluxes as the outcome of the LCC.

3.5.1 Deforestation and Forest Degradation

Forest degradation refers to the reduction in forest biomass due to fire or anthropogenic disturbances like non-sustainable harvest, selective logging, or fuel-wood collection. Deforestation refers to the removal of forest to use the cleared land for other purposes, typically agriculture. Deforestation occurs worldwide at various extent and rate. In 2010, deforestation was the highest in the Tropics of America, Africa, and Asia with $4.5 \cdot 10^4\,km^2/year$, $3.1 \cdot 10^4\,km^2/year$, and $2.9 \cdot 10^4\,km^2/y$, respectively (`http://en.wikipedia.org`).

Most of our knowledge on the climate impacts of deforestation bases on numerical modeling studies. In response to deforestation, surface-roughness length decreases, and albedo and emissivity change (cf. Table 3.1). Deforestation reduces the infiltration capacity due to soil compaction during and after the deforestation. The comparatively smaller root-zone of short-vegetation than forest restricts the plant-available water that can support evapotranspiration. Typically, deforested areas show a greater spatial variability of soil moisture than prior to deforestation. The deforestation namely enhances surface runoff that redistributes the received rainfall. Deforestation also yields to increased stomatal resistance (cf. Table 3.1), which may contribute to the lengthening of the Amazon dry season (D'Almeida et al. 2007).

The biophysical effects of and atmospheric responses to deforestations differ strongly with latitude. In seasonally snow-covered high latitudes, deforestation leads to cooling due to the increased surface albedo in winter (Li and Mölders 2008). In the Tropics, deforestation reduces evaporation and the associated evaporative cooling. Consequently, tropical deforestation fosters warming, while boreal deforestation opposes warming. The evaporative impact of temperate forests and the net effect of its deforestation on climate are still under research (Bonan 2008). The response of precipitation to boreal and tropical deforestation differ because of the implications that baroclinic cold and warm frontal dynamics have for precipitation formation at high latitudes in summer (Pielke et al. 2007).

3.5.1.1 Tropical Forests

In the early- and mid-1970s, Amazonia experienced LCC because of the enhanced efforts to develop the region economically. Due to this development, more than 10% of Amazonia's original forest was replaced by pasture, cropland, or soybean agriculture by the 1990s. The worldwide public concern about the LCC in Amazonia – Amazonia holds more than

40% of the Earth's remaining tropical rainforests – led to many research studies with focus on the climate impacts of tropical deforestation.

Results from field campaigns like ARME (e.g., Shuttleworth 1988; Wright et al. 1992) led to modeling studies at various scales assuming different tropical deforestation scenarios. D'Almeida et al. (2007) reviewed the sometimes seemingly contra-dictionary results of these modeling studies. Obviously, the magnitude of changes in response to deforestation is strongly sensitive to the assumed new land-cover, the horizontal extent of deforestation, and the scale at which the climate impacts were investigated.

The results of GCMs generally agreed that tropical deforestation on the long term decreases precipitation and evapotranspiration, and increases near-surface temperature (D'Almeida et al. 2007). Climate model simulations indicated that even an increase as low as 0.03 in albedo over the Amazon rainforest due to deforestation may be sufficient to decrease rainfall over Amazonia (Nicholson et al. 1998). In response to the deforestation, evapotranspiration goes down due to the reduced surface-roughness length and statistically significant decreased net radiation. The reduced evapotranspiration subsequently modifies the water and energy cycles within the deforested region and weakens the large-scale atmospheric circulation.

Since runoff contains uncertainty from precipitation, infiltration, and evapotranspiration, simulated runoff trends differ largely among GCMs. However, many GCMs indicated a long-term decrease in runoff due to deforestation. The magnitude of decrease depends strongly on the projected precipitation change that is expected to exceed the change in evapotranspiration at the basin-scale (D'Almeida et al. 2007). The increased Bowen ratio observed after deforestation yields to increased near-surface temperature. Modeling studies by Henderson-Sellers et al. (1993), for instance, showed that deforestation of the Amazon rainforest reduces temperature and increases precipitation south of the deforested area.

The GCM simulations on tropical deforestations also indicate significant climatic changes in areas far away from the actual disturbance due to teleconnection (Werth and Avissar 2002). Various modeling studies suggest that the Amazon rainforest contributes to sustaining the Hadley and Walker circulations. Tropical deforestation affects the Hadley and Walker circulations by weakening of the circulation (Shukla et al. 1990; Henderson-Sellers et al. 1993). The reduced evapotranspiration and upward motion over land after deforestation alter the surface heat and moisture fluxes. Since these fluxes and cumulus convection are linked with each other, deforestation alters the spatial distribution of deep cumulus convection and the release of latent heat from condensation. Deforestation reduces the release of latent heat from deep cumulus convection at the mid- and upper levels of the troposphere. Thus, deforestation weakens the tropical Hadley circulation and thereby modifies the Rossby-wave source. Consequently, changes occur in mid- and high-latitude circulations (Jonko et al. 2010) leading to changes in climate there.

Studies by Avissar and Werth (2005), for instance, indicate that deforestation of Amazonia severely decreases spring rainfall in the southern Midwest and winter rainfall in the northern Midwest. Deforestation of Central Africa reduces summer rainfall in the southern Midwest and spring rainfall in the northern Midwest. Deforestation in Southeast Asia affects precipitation most significantly in China and on the Balkan Peninsula. Deforestation of any of these tropical forests increases appreciably the summer rainfall in the southern part of the Arabian Peninsula. This region experiences the largest increase in annual accumulated precipitation for concurrent deforestation in all three tropical regions. Concurrent deforestation in all three tropical regions also decreases winter precipitation significantly in California. These results base on twelve 12-year long simulations performed with the Goddard Institute for Space Studies Model II global climate model. The six control simulations assumed the Amazon rainforest extent of 1950, whereas the six deforestation simulations assumed shrubs and grassland instead of rainforest from the start ($t = 0$). Analysis of the ensembles moreover showed that deforestation in Amazonia reduces rain-season precipitation in various regions of the world notably. These changes correlate well with the changes in Amazonia. To test the significance of the deforestation signal, the authors created three "false" control and three "false" deforestation ensembles wherein they combined the control and deforested members randomly. Comparison of the false ensembles with the actual "true" ensembles revealed nonrandom variation in precipitation. The globally averaged precipitation deficits for the true ensembles generally exceeded those of the false ensembles notably.

The changes in evapotranspiration and sensible heat flux in response to deforestation affect the dynamic structure of the ABL (Sivakumar 2007). In the Tropics, NCMC can be observed, especially during the dry season because then soil moisture differs greatly between the deforested areas and the natural forest. Such NCMC may lead to the different responses of the water cycle to deforestation obtained by mesoscale models and GCMs for the dry season for the same region. During the rain season, the strong synoptic scale forcing suppresses the formation of NCMC

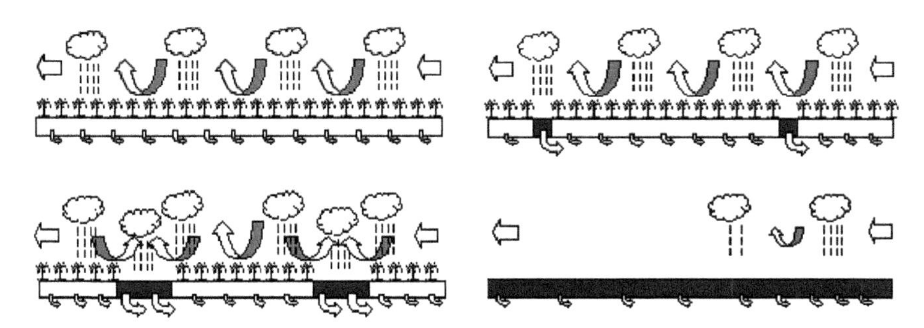

Figure 3.7. Schematic view of the impacts of various extents of clearing (*dark gray*) in Amazonia on the atmosphere. Precipitation recycling sustains the horizontal water-vapor flux that transports moisture into the region. This flux maintains high rainfall amounts in the case of no deforestation (*upper left*). Local deforestation (<100 km^2) is too small to affect rainfall, but increases runoff and decreases evapotranspitation (*upper right*). Regional deforestation (100–10,000 km^2) influences the circulation, strengthens convection, and potentially increases rainfall (*lower left*). Increased soil moisture delays while increased atmospheric relative humidity accelerates the onset of rainfall. However, both increase rainfall. Basin-wide deforestation (>10,000 km^2) would reduce evapotranspiration and precipitation recycling severely (*lower right*) (Modified after D'Almeida et al. 2007)

so that tropical deforestation cannot significantly affect the cloud and precipitation distribution (Sivakumar 2007).

Satellite observations show increased cloudiness at the border between deforested and forest patches during the dry and rain season (Ramos Da Silva et al. 2008). The local occurrence of convective clouds permits relatively more solar radiation to reach the surface than it does under stratiform cloud conditions. Consequently, the surface sensible heat fluxes increase, which reinforces the atmospheric instability and convection (Ramos Da Silva and Avissar 2006).

The pattern and degree of deforestation determine the atmospheric response to deforestation strongly (Fig. 3.7). High-resolution mesoscale simulations and satellite images show that the typical fishbone pattern of Amazonia deforestation (Fig. 3.1) yields coherent NCMC (Baidya Roy and Avissar 2002). The Rondonia Boundary Layer Experiment (RBLE-3) in Brazil showed that shallow (deep) convection tends to occur over deforested (forest) areas. Atmospheric soundings performed simultaneously at forest and pasture sites indicated that the ABL is more unstable and has larger CAPE over the forest than deforested areas. Nonclassical mesoscale circulations yield to the relatively stronger activity of shallow convection over the deforested areas than deep convection over forest. The synoptic flow transports these perturbation signatures away, and the moisture and temperature perturbations modify the synoptic scale moisture and heat transport with consequence for climate.

High-resolution mesoscale model simulations showed that in Amazonia, complete deforestation at the benefit of pasture leads to higher surface sensible fluxes than the typical fishbone deforestation pattern (Fig. 3.1) or a forest-only landscape. Consequently, circulations are more intense in the pasture-only landscape than in the other cases. Evapotranspiration is largest and lowest for the forest-only scenario and pasture-only scenario, respectively (D'Almeida et al. 2007).

Observations on the local scale indicate that evapotranspiration within small deforested areas depends on local land-surface characteristics rather than the synoptic forcing. At local scale, the natural variability induced locally exceeds the magnitude of signals induced globally (D'Almeida et al. 2007).

3.5.1.2 Temperate Forest

In midlatitudes, deciduous forest has been converted frequently to grasslands and pastures, particularly in regions with generally poor soil conditions. Such LCC reduce the spring-snowmelt rates, transpiration, LAI, precipitation interception, moisture availability, and the height of LCL, and enhance evaporation of soil water and runoff. Enhanced runoff may raise the frequency of flooding and/or soil erosion (Pielke et al. 2007). The change in soil moisture due to replacement of temperate forests by grassland and/or cropland decreases regional precipitation (Mölders 2000a). The impact of replacing temperate forests by agriculturally used land on the regional moisture fluxes depends on irrigation practice and the water demands and growth cycle of the crop grown.

The net climate impact of temperate forest loss is still under research (Bonan 2008). Effects from evapotranspiration during summer and increased albedo during winter compete for their influence on annual mean temperature. The increased albedo due to deforestation could offset carbon emission, leading to a negligible net climatic effect of temperate deforestation. The reduced evapotranspiration due to deforestation could amplify biogeochemical warming (Bonan 2008).

3.5.1.3 Boreal Forest

The atmospheric impacts of coniferous and deciduous boreal forests differ strongly. Summertime evaporative fraction (defined as the ratio of latent heat flux to available energy) of deciduous broadleaf forests exceeds that of coniferous forest. Consequently, the sensible heat exchange rates and ABL heights over coniferous forest exceed those over deciduous forest.

Deforestation in boreal regions occurs mostly due to cuttings and wildfires. Wildfires prefer coniferous over deciduous forest. Both kinds of deforestation change the radiative (albedo, emissivity), vegetative (vegetation type, fraction, LAI, stomatal resistance), thermal (soil heating, increase of active layer depth), and hydrological (water loss, reduced transpiration) surface properties. In boreal forests, wildfires lead to reduced albedo in the snow-free season and enhanced albedo in the first winters after the fire. The fires remove the insulating moss and/or peat layers and expose the underlying mineral soil that has different thermal and hydrological regimes than the original organic soil. The heat generated by the fires is conducted to the permafrost (i.e., soil in which temperatures remain below 0°C for at least two consecutive years) and thaws some of the permafrost. Thus, the thickness of the active layer (i.e., the layer of soil that overlays the permafrost and thaws seasonally) increases. Consequently, a deeper soil layer is available to provide plant-available water for evapotranspiration during summer.

In AVHRR imagery, intensely burned areas show up to 6 K higher skin temperatures even many years after the fire (Amiro et al. 1999). According to aircraft measurements, sensible heat fluxes into the atmosphere are about 10–20% higher over burned areas than their surroundings for the first few years after the fire, while latent heat fluxes are slightly lower. The Bowen ratio increases about 50% as compared to the undisturbed surroundings. The partitioning of the surface energy and water fluxes, however, affects strongly the water and heat exchange with the atmosphere. Observations show reduced CO_2-fluxes for about 15 years after the fire, with the largest reduction (up to 25%) in the year after the fire (Amiro et al. 1999).

In boreal forest, the heterogeneity influences the likelihood of wildfires (Hély et al. 2001). Wildfire size correlates positively with the abundance of pine forest in the vicinity and negatively with the presence of recently burned areas. The presence of fine fuels like grass and shrubs may increase the total burned area because these plants regenerate faster than heavy fuels like trees. This means the presence of fine fuels shortens the intervals between wildfires.

Wildfire-induced LCC affect the local climate by altering the fraction of solar radiation backscattered to space and shifting the partitioning of net radiation between ground, sensible, and latent heat fluxes in favor of enhanced sensible heat fluxes. Wildfire-induced LCC influence the regional climate by modifying the ABL structure, the conditions for thermally induced convection, the dominant paths for cloud and precipitation formation processes, and mesoscale circulation pattern by nonlinear feedback mechanisms with potential implications for climate variability.

Mölders and Kramm (2007) performed simulations with the mesoscale meteorological model generation 5 (MM5) with same meteorological initial and boundary condition for the landscape of Interior Alaska prior and after the 2004 fire season which was the strongest since onset of recording. Even though the change in sensible heat fluxes due to the burned area may be larger for small than large burned areas, a burned area must exceed 600 km^2 to heat the air long enough to initiate notable upward motions at its downwind end and cause significant changes in cloud and hydrometeor mixing ratios and precipitation. To produce significant changes in the latent heat fluxes and vertical wind speed, the burned area must exceed 800 and 1,600 km^2, respectively.

The increased lifting of warm and moist air over young fire scars shifts the cloud and precipitation-formation processes toward the cold, more efficient path leading to enhanced precipitation in the lee of the fire scar. Behind this area of enhanced precipitation, an area of reduced precipitation occurs because more water vapor was removed from the atmosphere in the upwind earlier (Fig. 3.8). Mölders and Kramm (2007) found that the differences in density, pressure, temperature, and enhanced upward motions can induce NCMCs if the burned area is large enough (Fig. 3.9).

The authors' simulations indicated that fire scars of sufficient extent influence significantly the temporal and spatial precipitation distribution without notable changes in the landscape-average precipitation. Local increases in precipitation rates exceed frequently 12%, thereby increasing the risk of flooding by creeks.

Boreal forests experience 6–8 months long snow season. As snow falls on deforested areas or young fire scars, it stays on the ground and albedo increases drastically within hours. On the contrary, boreal forest intercepts some of the snow. Wind and/or the snow load partially or totally empty the interception storage, thereby reducing albedo. Thus, snow-covered fire scars or deforested areas reflect more incoming shortwave radiation to space than boreal forest does and thereby cool the atmosphere. The lower temperatures may feedback to more solid precipitation, which refreshes the high-surface albedo. If deforested areas or fire scars are on mountaintops or plateaus, cold air pools will build that

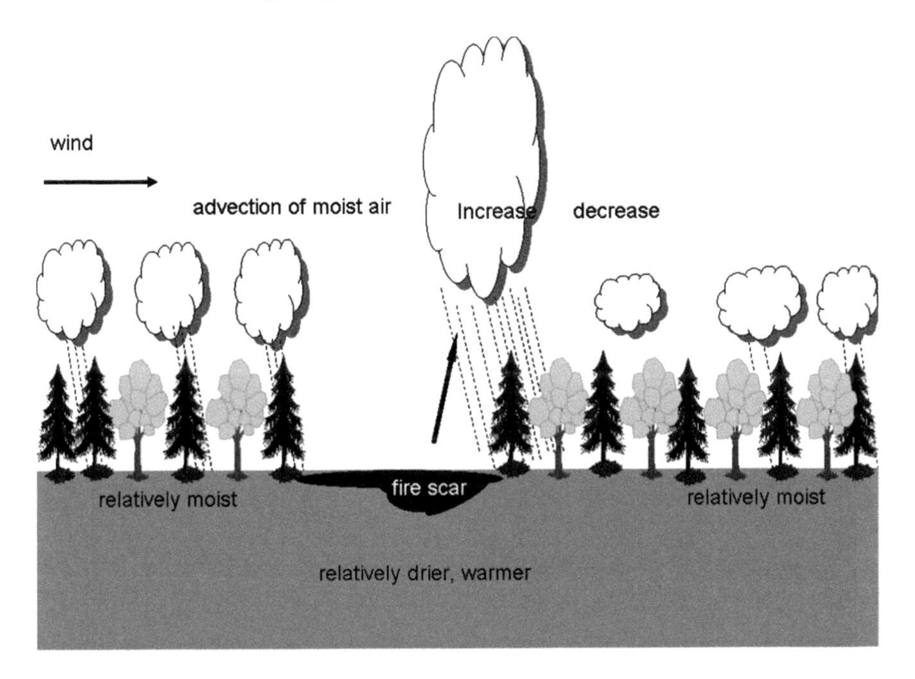

Figure 3.8. Schematic view showing the modification of atmospheric processes in response to boreal forest loss due to wildfire (From Mölders and Kramm 2007)

may drain into the valleys. Here, the cold air may strengthen inversions, which may worsen air pollution in populated high-latitude valleys.

3.5.1.4 Trace Gas Impacts

Forests make up nearly 50% of the global terrestrial carbon pool, and about 75% of the living carbon if soils are excluded (Corbera et al. 2010). Thus, forest degradation or changes in forest cover not only alter the physical surface properties, but also the carbon budget (Bonan 2008) and the trace gas cycle. Deforestation and forest degradation reduce substantially the forest-carbon stocks and can affect GHG emissions. The conversion of tropical forest to agriculture, for instance, contributes to an increase of CO_2 in the atmosphere. In general, deforestation may have potentially a warming effect on the atmosphere as it eliminates the potential to store CO_2 in trees, and as it will release CO_2 if performed by fire. These effects are called "carbon cycle effects" (e.g., Bonan 2008).

Due to the lack of standardized methods, the estimates of emissions from deforestation vary widely between 0.8 and 24 Gt/year. Emissions of

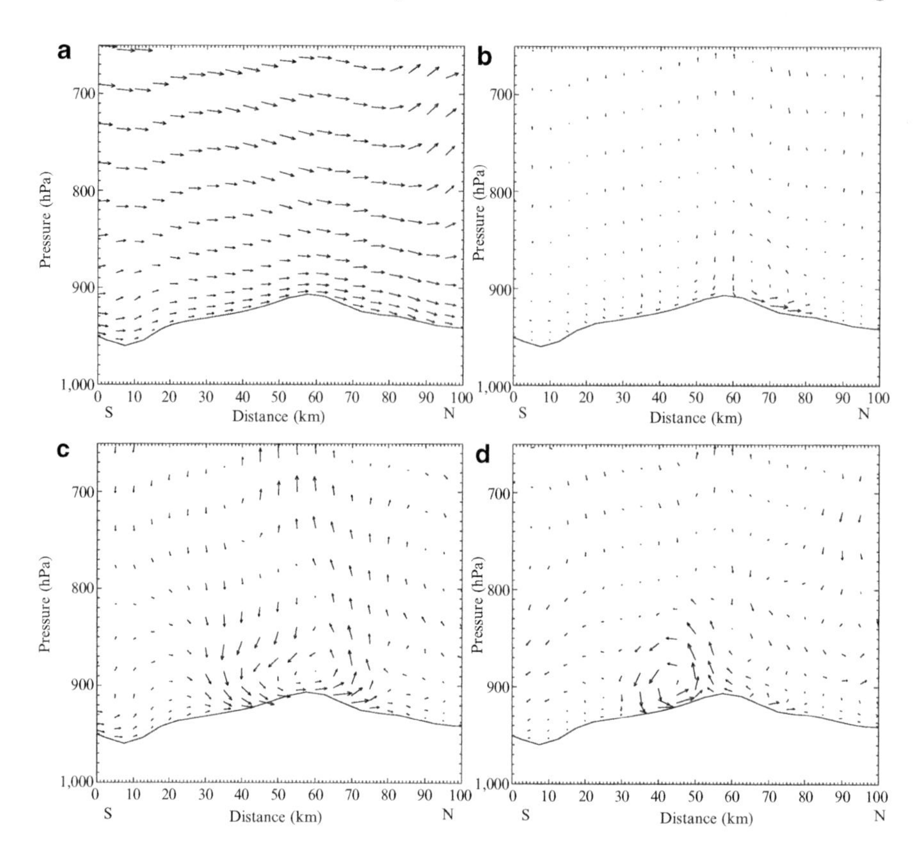

Figure 3.9. S-N cross-section over a burned area located between 25 and 75 km. (**a**) Wind vectors as obtained by the simulation with the landscape prior to the wildfires at 0500 Alaska Standard Time (*AST*). The maximum horizontal ($v_{\max}$) and vertical motion ($\omega_{\max}$) are 7.3 m/s and 89 Pa/s, respectively. Differences of wind vectors obtained from the simulations with the landscape after and prior to the wildfires at (**b**) 0500 AST ($\Delta v_{\max} = 0.5$ m/s, $\Delta\omega_{\max} = 5$ Pa/s), (**c**) 1100 AST ($\Delta v_{\max} = 0.2$ m/s, $\Delta\omega_{\max} = 8$ Pa/s), (**d**) 1700 AST ($\Delta v_{\max} = 0.3$ m/s, $\Delta\omega_{\max} = 9$ Pa/s). Note that the synoptic-scale flow remains nearly constant for the entire episode (From Mölders and Kramm 2007)

GHG from LCC increased by 40% from 1970 to 2004. Emission of CH_4, nitrous oxide (N_2O), and other chemically reactive gases from the land-cover subsequent to deforestation contributed about 20–25% of the total anthropogenic GHG emissions during the 1990s (Corbera et al. 2010).

Volatile organic compounds (VOC) include organic atmospheric trace gases other than CO_2 and carbon monoxide (CO) like acids, alcohols, alkanes, alkenes, carbonyls, esters, ethers, and isoprenoids (isoprene, monoterpenes). VOC can influence significantly the tropospheric chemistry as they control the processes of atmospheric haze and ozone formation (Guenther 1997). Biogenic emissions of VOC depend strongly on PAR, leaf-age, phenological events, foliage temperature and density, and vegetation species. A mixed forest with 20% hickory and 80% oak trees, for instance, has an average potential VOC emission that is a factor of four higher than that of a forest with 80% hickory and 20% oak trees. At standard conditions of temperature (30°C) and PAR (1,000 μmol $m^{-2}\,s^{-1}$), forest can produce more than double the isoprene emissions of grassland. Isoprene emissions from forest are a factor 100 to 30% lower than for various croplands. Monoterpene emissions from forests can exceed those of various grasslands between about 10% and up to 7.5 times and those of croplands between about 6.5 and 21 times (Guenther 1997). Crops, except sunflowers, are among the low monoterpene-emitting species. Wheat, rice, rye, sorghum, beans, and soybeans emit hardly any isoprenes or monoterpenes. White spruce, Sitka spruce, and black spruce emit up to more than a factor two more isoprene than alder; these spruce species emit monoterpenes in similar order of magnitude than birch do (Kesselmeier and Staudt 1999).

Birch and alder are among the first trees to grow after boreal forest fires. These great differences in foliar emissions of VOC, especially isoprenes and monoterpenes, between the various forest successor land-covers and forests make deforestation and/or forest degradation a strong modifying factor of atmospheric composition and chemistry. The fate of organic atmospheric carbon and its oxidation products and/or transformation to aerosols is a hot research topic.

3.5.2 Conversion of Native Grasslands to Cropland

The conversion from native midlatitude vegetation to agriculturally used land that has been studied the most intensively in the context of the settling of America (e.g., Copeland et al. 1996; Chase et al. 1999; Pielke et al. 2007). Prior to the settling of North America, large areas were short-grass steppe dominated by low stature (5–30 cm), drought tolerant, warm-season grasses. Today, a mosaic of dryland, irrigated croplands, and native vegetation covers this region (e.g., Fig. 3.10). The irrigation increases evapotranspiration as compared to the original short-grass steppe. Short-grass steppe can be photosynthetically active year-round as long as atmospheric conditions permit. The peak in biomass production and water demands of short-grass steppe differ

Figure 3.10. Aerial photo of a mosaic of natural, irrigated, and nonirrigated land (From `http://permaculture.org.au/2009/11/02/rethinking-water-a-permaculture-tour-of-the-inland-northwest/`)

strongly from those of cropland. Cropland is bare after the harvest meaning a distinct change to the former seasonality of water supply to the atmosphere. Like for deforestation, LCC from other native vegetation to agricultural (or urban) land decrease evapotranspiration and increase runoff (Huntington 2006).

Another important impact on the energy budget results from the different temporal evolution of albedo. Cropland and grassland have similar albedo (Table 3.1) during the crop-growing season. After the harvest, however, the albedo of the now exposed bare soil is lower than that of the original short-grass steppe.

Surface roughness of agricultural plants exceeds that of short-grass steppe (Table 3.1). This means as compared to the native vegetation the seasonality in sensible and latent heat fluxes, albedo, LAI, and surface roughness changed due to the LCC. At local scale, temporally increased humidity and decreased temperatures as compared to the conditions over the native vegetation exist over irrigated areas during the growing season (Chase et al. 1999). The impacts on temperature, atmospheric humidity, and precipitation vary with location, extent of the patches of same land-cover, and degree and type of LCC. Locally, temperatures increased up to 3 K in areas converted to nonirrigated cropland (Pielke et al. 2007).

Almost 100% of the native tall-grass steppe has been converted for agricultural use worldwide. This LCC led to similar impacts as described for the conversion of short-grass steppe except for the loss of biomass and increase in LAI.

Middle and Eastern Europe have a long history of draining wetlands and marshes for agricultural use. These draining efforts reduced the heterogeneity of the landscape and local rainfall. Mölders (1999b) used a mesoscale meteorological model to examine the atmospheric influences of the drainage projects that occurred 200 years ago in East Germany. In summer in the afternoon, conversion of marshland to grassland reduced the sensible heat-fluxes about $100\,\mathrm{W\ m^{-2}}$. The opposite is true in the morning. Around noon, latent heat fluxes are locally up to $50\,\mathrm{W\ m^{-2}}$ lower and more than $100\,\mathrm{W\ m^{-2}}$ higher over the now grassland than the former marshland. The horizontal extent of LCC and the new land-cover (grassland or cropland) play a role. The nonunanimous response of latent heat fluxes also partly results from the reduced insolation due to enhanced cloudiness. On average, cloudiness is less in the original than drained landscape in the morning and early afternoon, while later the opposite is true. The lower evapotranspiration over the marshland converted to grassland reduces near-surface humidity by more than 0.5 g/kg at noon. Typically, the draining shifts the partitioning of energy toward higher Bowen ratios than in the original landscape. On average over the landscape, cultivation of marshland reduces evapotranspiration and leads to drier atmospheric conditions.

The lower soil-heat capacity, albedo, and higher thermal conductivity of the drained than non-drained soils reduce near-surface air temperatures up to 2 and 0.5 K on average at noon. In the upper ABL, air temperatures are up to 0.4 K higher than over the former marshland. Thus, drainage of marshland increases the atmospheric stability (Mölders 1999b).

In the non-drained landscape, cloud fields form over the area of the lowest Bowen ratios after noon. Secondary changes (modified cloud processes, dynamics, insolation, etc.) resulting from the primary responses (modified thermal stratification, evapotranspiration, etc.) cause differences in cloud distributions. The changes in cloudiness result mainly from changes in vertical motions due to the different surface heating and evapotranspiration. In the original landscape, precipitation sets on later and more rainfall occurs at fewer places than in the drained landscape. Draining homogenizes precipitation, shifts the position of accumulated rainfall maxima slightly and reduces maximum precipitation (Mölders 1999b).

3.5.2.1 Impact of Irrigation

Worldwide, the irrigated area increased 50-fold since 1970 (Sivakumar 2007). Irrigation repartitions the sensible and latent heat fluxes during and after the event as compared to nonirrigated or native conditions. The increased surface wetness enhances the latent heat fluxes and decreases the sensible heat fluxes. Due to irrigation dewpoint, temperatures and moisture fluxes increase in the ABL, for which CAPE, atmospheric instability, and daytime cloud-cover may increase.

Changes in moisture-flux convergence and CAPE that reinforce mutually modify precipitation coherently (Saulo et al. 2010). Various studies indicate that irrigation leads to increased precipitation under synoptic-scale condition with low-level convergence and uplift. Under these conditions, the additional moisture from irrigation often permits air parcels to reach the saturation level. The enhanced moist static energy may result in increased rainfall (Pielke et al. 2007).

The enhanced evaporation cools the surface layer. This cooling associated with irrigation led to the hypothesis that the increased irrigation has “masked” the global warming signal (Kueppers et al. 2008). Model simulations performed alternatively with irrigation of cropland and with the original landscape of Nebraska, for instance, indicate that irrigation reduces near-surface air temperatures by about 3 K and increases the surface latent heat fluxes by 42% on domain-average (Pielke et al. 2007).

Irrigation in great style may affect mesoscale circulation patterns. Observational studies in eastern India, for instance, indicate that irrigation can decrease sea-breeze convection and regional rainfall during the pre-summer monsoon season (Pielke et al. 2007).

3.5.2.2 Trace Gases

Conversion of grassland to agriculturally used land affects the emission and uptake of species and hence GHG. Cropping management affects strongly the GHG emissions associated with LCC (Kim et al. 2009). Agriculture contributes to CO_2 emissions directly due to the use of fossil fuels in the production and indirectly due to the use of energy needed for fertilizer production.

Methane is the simplest alkane, and a relative potent GHG. It is primarily removed by reaction with hydroxyl radicals in the atmosphere, producing CO_2 and water. Once CH_4 reaches the stratosphere, it contributes to the destruction of the ozone layer. Wetlands, rice paddies, and the ruminant, grazing herbivores (cattle, sheep, goats) on rangeland emit CH_4. Land-cover related CH_4 sources are anaerobic production by methanogenesis and decomposition of biomass. Obviously, plants absorb CH_4 from the soil and emit it through their leaf tissue. Whether some plants produce CH_4 is a hot research topic (Keppler et al. 2006).

Agriculturally used land contributes notably to the atmospheric N_2O concentrations, while grasslands are minor contributors in extensive grazing regimes. Soils emit NO due to nitrification and denitrification processes. Emission of NO from agriculturally used fields strongly depends on soil temperature and moisture, land management (kind of fertilizer, fertilization rates), the cultivated species and density of cultivation (Fang and Mu 2007). NO emission from tilled cornfields, for instance, can exceed those from the untilled cornfields by about five times; fallow land with sparse vegetation emits NO at higher rates than land with a closed canopy. Emission rates of unfertilized fallow land can exceed those of fertilized agricultural cultivations with closed canopy. Wheat, soybeans, cabbage, and celery fields are usually sinks for atmospheric NO_2, while potato fields and, at low ambient NO_2 concentrations, orchards can be sources (Fang and Mu 2007).

Conversion from grassland to agriculturally used land affects the VOC emissions. Grassland emits three orders of magnitude less methanol than agricultural crops. Beans rice, rye, sorghum, soybeans and wheat have negligible isoprene and monoterpene emissions, while alfalfa and cotton emit noteworthy amounts of monoterpenes. Clover emits about 30% less ethane, ethane, nonanal, and n-pentane, eight times less ethanol, six times less acetaldehyde, ten times less hexanal, and 50% less pentanal than grass. However, clover emits 4.2, 12.6, and 86 times more acetone, methanol, and butanone than grass (Kesselmeier and Staudt 1999).

3.5.3 Droughts and Desertification

Various definitions for desertification exist. Some scientists use the word "desertification" in the sense of a gradual change in progress, while others use it as the end product of a desert-like status quo. All definitions have in common that desertification refers to land degradation in dryland of arid, semiarid, and dry subhumid regions. Loss of vegetation cover is a major reason of land degradation (Sivakumar 2007).

Dryland exists in about 34.4%, 24.1%, 24%, 15%, and 2.5% of Asia, Africa, Americas, Australia, and Europe, respectively. Dryland is characterized by its defined climatically low ratio of long-term average annual precipitation to potential evapotranspiration. Rainfall shows a daytime dominance. The absorption of solar radiation by the surface and surface-to-atmospheric heat fluxes drive the convection and determine the ABL characteristics. If enough moisture exists, clouds and rainfall may occur (Sivakumar 2007).

Climate and desertification interact in a two-way manner. Typically, the impacts of desertification on climate are described in terms of LCC that caused land degradation, emissions, and pollution due to biomass burning, accelerated wind erosion, and salinization of soils due to irrigation (Sivakumar 2007). However, desertification can result from anthropogenic activities (e.g., over-grazing, deforestation, changes in agricultural practices), climatic variations (e.g., drought), and/or changes in ocean-temperature distributions (Giannini et al. 2008).

Discussion on desertification due to excessive use of wood fuel, overgrazing, and overcultivation started during the Dust-Bowl time in the 1930s. During that time, a strong drought in the Great Plains ruined farmland and led to soil erosion. Later research using GCMs indicated that teleconnection contributed to the drought, Dai (2011). Cooler and warmer than normal SST in the tropical Pacific Ocean and tropical Atlantic Ocean, respectively, created shifts in weather patterns and the jet stream position. Less moisture was advected from the Gulf of Mexico leading to reduced rainfall throughout the US Midwest.

The intense Sahel drought of 1968–1973 brought desertification and the potential impact of land-cover on weather again to worldwide attention. Results from different climate models suggest that the persistence of the 1968–1973 Sahel drought was caused by ocean, not land processes (Giannini et al. 2008). The drought is statistically and dynamically associated with enhanced SST in the tropical oceans, especially the Pacific and Indian Oceans, and concurrent increased warming of the southern hemispheric relative to the northern hemispheric Atlantic. Observations also showed the progressive pattern from the wetter-than-normal 1950s and 1960s to the drier-than-normal 1970s and 1980s that correlates well with the warming of the global tropical oceans (Giannini et al. 2008). Note that the 1980s were the driest decade in the Sahel in the last century.

Simulations performed with ECHAM showed that a 1K cold anomaly of the Indian Ocean leads to positive rainfall anomalies in the Sahel west of 20°E. Superimposing the positive SST anomalies in the Indian Ocean observed in the 1980s led to drying. The westward-propagating Rossby-waves establish the atmospheric teleconnection between the Indian Ocean and Sahel where they induce subsidence. The partial recovery of the rains in the 1990s agrees with the competing influences of Indian and Atlantic SST. While the Atlantic and Indian Ocean dominate the climate of the Sahel on the long time scale, the Pacific modulates the Sahel climate through the El Niño-Southern Oscillation that occurs on temporal scales of 2–7 years. The El Niño and La Niña events correlate with drought and abundant rainfall in the Sahel, respectively (Giannini et al. 2008).

These findings indicate that the oceans force the climate variability of the Sahel at interannual to interdecadal time scales. However, they fail to explain the full strength of the droughts. Obviously, the land-surface is a secondary factor and can enhance locally the remote forcing from the oceans. Potential mechanisms involve soil moisture, dust, and vegetation (Giannini et al. 2008).

According to Charney et al.'s (1975) biogeosphere–albedo feedback theory, reduced vegetation may feedback to reduced rainfall (positive feedback). The increased albedo due to reduced vegetation cover enhances radiative cooling. The enhanced descent, reduced rainfall and consequently further reduction of vegetation balance the enhanced radiative cooling.

Various GCM simulations provide evidence for this hypothesis (Nicholson et al. 1998). Increasing the surface albedo of Sahel vegetation from 0.2 to 0.3 reduces moisture convergence and rainfall in the Sahel and south of it. The rainfall anomaly and changes in circulation agree well with those observed during the Sahel drought. GCM simulations assuming a doubling of the global extent of deserts showed notable impacts on Sahel rainfall. This means that in drylands, local land-management practices can lead toward loss of productivity and degradation (Sivakumar 2007).

The increased cloudiness required for increased rainfall may reduce incoming solar radiation, thereby reducing evapotranspiration. This cloud-evaporation feedback means a negative feedback on rainfall. Clouds also reduce the long-wave radiation emitted from the Earth, thereby affecting the energy budget and temperature.

The amount of plant-available water in the root-zone, and not the absolute amount of precipitation, is essential for plant survival. Thus, droughts result in sparse vegetation cover, except where drainage lines concentrate water or the groundwater level is high, that is in reach of the roots. The high daytime temperatures hinder the storage of organic matter and hence nitrogen in the soil (Verstraete et al. 2009). Desertification reduces or even eliminates carbon sequestration from vegetation and can destabilize regional climate, ecosystems, and hydrological systems.

In degraded areas, the sparse vegetation cannot prevent soil-uptake by wind, especially during severe weather. Dust uptake means loss of soil and further reduces plant productivity. The airborne dust affects the radiative budget directly through reflection and absorption of solar radiation and indirectly through modification of cloud optical properties and lifetime. The reduced insolation and energy available for evapotranspiration feed back positively to less rainfall. Since some dust particles may act as CCNs, more dust may reduce the efficiency of precipitation formation that also feeds back to less rainfall.

If the ocean conditions cause a rainfall anomaly, the surface processes can amplify the land–atmosphere response if the positive feedbacks dominate. If the negative feedbacks dominate, an intact land-cover can mitigate the oceanic impact on the Sahel's climate (Giannini et al. 2008).

Beside the loss of vegetation, salinization increases the albedo locally up to 0.6 where saltpans and dry salt lakes form. Here, surface temperatures are up to 16 K lower than over the surrounding bare or sparsely vegetated land during the day; at night, temperature are about 8 K higher over the salt lake than the surroundings (Sivakumar 2007). These surface-temperature gradients may induce NCMC. The daytime divergence and nocturnal convergence can enhance wind erosion and dust deflation further.

Observations suggest that soil moisture and land–atmosphere coupling affects the observed 15-day westward-propagating mode of intra-seasonal variability in the West African monsoon. Lavender et al. (2010) investigated this hypothesis with a set of three atmospheric GCM simulations. The GCM simulations with fully coupled soil moisture show the 15-day mode of land–atmosphere variability clearly. Precipitation anomalies result into soil-moisture anomalies within 1–2 days; satellite observations confirm these findings. A set of their simulations examined whether soil moisture affects precipitation merely passively, or participates actively in this mode. Simulations in the fully coupled mode with a forced 15-day westward-propagating cycle of regional soil moisture anomalies showed the reduced surface sensible heat flux over the imposed wet-soil anomalies leads to negative low-level temperature anomalies and enhanced pressure. The anti-cyclonic circulation around these cool high-pressure regions enhances the northward moisture advection and precipitation westward. The soil moisture–forced precipitation response provides a self-consistent positive feedback on the westward-propagating soil-moisture anomaly. This response indicates an active influence of soil moisture on the precipitation regime. In another sensitivity simulation, the authors prescribed soil moisture again externally, but suppressed all intra-seasonal fluctuations. Even without soil-moisture variability, surface sensible heat fluxes vary strongly due to cloud changes. This variability in surface sensible heat fluxes produces the 15-day westward-propagating precipitation signal, but less coherent than in the simulations wherein the soil-moisture–atmosphere interaction was considered. Sensitivity studies revealed that the precipitation mode becomes phase-locked to the forced soil-moisture anomalies. This means the 15-day westward-propagating mode in the West African monsoon can occur independently of soil moisture. The soil-moisture–atmosphere coupling feeds back to the atmosphere, and enhances and organizes the 15-day mode (Lavender et al. 2010).

Climate variability can affect the location and strength of desertification. However, even though desertification occurs locally, it influences the atmosphere on the local, regional, and global scale. GCM simulations showed that in most of the Earth's deserts, decreases in evapotranspiration correlate positively with reduced precipitation. Some of the desert regions show decreased soil moisture and latent heat fluxes and increased near-surface air temperature and likelihood of dust uptake. Once dust is mixed upward into the free atmosphere, it becomes subject to long-range transport. Thus, due to the radiative effect of dust particles, desertification may affect regional climate far remote from where desertification occurred. Asian dust from the Gobi desert, for instance, is observed frequently in Alaska in March (Sassen 2005).

Like found for Amazonian deforestation that affects significantly the Walker- and Hadley-circulation and modifies climate far way from the region of deforestation by Rossby-wave propagation mechanisms, desertification can affect large-scale circulation patterns. Desertification of the former Mongolian grassland, for instance, weakens the monsoon circulation and reduces convective latent heating (Xue 1996). Historic records indicate north and southward movements of the Mongolian mixed grassland/agricultural belt for the last 2,000 years. In this region, the monsoon governs summer precipitation. Since the 1960s, intensive grazing desertified the Mongolian grasslands. Over this episode, observations showed a warming of the Inner Mongolia grassland and northern China, and an increase in precipitation over China and Inner Mongolia, but a decrease over northern and southern China.

Xue (1996) performed ensembles of 90-days GCM simulations wherein the Inner Mongolia and areas of Mongolia were set to grassland and alternatively to desert. The differences in simulated precipitation pattern agreed well with the observed precipitation changes due to the observed desertification. The desertification weakened the monsoon circulation and caused changes outside the desertified region. The reduced convective latent heating above the surface layer enhanced the descending motion over the deserts and the areas adjacent to the South. Such changes south of the desertification area were also found in model simulations for the Sahel, but with less horizontal extent. Obviously, the degree of desertification, local topography, and dominant circulation pattern determine how far the impacts may reach outside the desertification region.

In Xue's (1996) Mongolian desertification experiments, the reduced evapotranspiration dominated the response of the surface energy budget. The reduction in rainfall in the inner desertification area was less than the reduction in evapotranspiration. The increased upward extent of the moisture flux failed to compensate for the decrease in evapotranspiration. The study also indicated that despite increased long-wave radiative loss at the top of the atmosphere in the desertification simulations, the decreased convective heating governs the reduction in diabatic heating (Xue 1996).

3.5.4 Recultivation

Recultivation refers to changing an area so that it regains the same characteristics as the landscape prior to the LCC. Frequent recultivation efforts are reforestation, afforestation, taking fallow land back into agricultural production, reversing desertification or returning filled open-pit mines back to their previous use. In the last decades of the last century, farm abandonment, reforestation, and intrusion of woody species from fire suppression increased the extension of temperate forest. The growing needs for food and biofuel may reverse these trends in the future (Bonan 2008).

3.5.4.1 Afforestation and Reforestation

Forestation refers to afforestation and reforestation. Afforestation is the planting of trees on areas where trees have never grown, to convert open land into forest or woodland. In the process of reforestation, native trees serve to restock the already existing depleted forests. While afforestation is an artificial, man-made process, reforestation can occur naturally. Nevertheless, both afforestation and reforestation can occur by seeding, plantation, or natural regeneration and create secondary forest. Secondary forest refers to forest that grows back after a disturbance.

The temporal evolution of impacts on the atmosphere varies among forestation processes. Generally, forestation decreases runoff and increases evapotranspiration (Huntington 2006). Natural regeneration and seeding have the longest recovery time. Seeding may involve later thinning of the stand. Such thinning again disturbs the system, that is increases runoff and decreases evapotranspiration relative to the conditions before thinning. Evapotranspiration (runoff) increases (decreases) much faster when planting than seeding the new stand. Planting leads to a very organized stand. Well-organized stands may develop internal circulation pattern later that had not existed in the original natural stand. Secondary forest leads to a quick recovery of evapotranspiration. Depending on the plant species, young secondary forests may provide similar evapotranspiration than mature forests (e.g., D'Almeida et al. 2007).

Typically, one speaks of primary and secondary forests also in the context of anthropogenic disturbance of primary forest (e.g., deforestation in Amazonia). About 30% of the accumulated deforested area in Amazonia is now secondary forest. Most of the boreal forest in a region with a wildfire-disturbance regime is secondary forest. In Alaska, where a wildfire-disturbance regime exists along a latitudinal corridor, the area of primary and secondary forest amount 82,350 and $222,999\,km^2$, respectively (Euskirchen et al. 2009).

After fires, if nature is kept alone, a successive development of landscapes will occur on a local to regional scale. As mentioned in the introduction, in boreal forest, for instance, succession landscapes have little to no growth in the year of the fire. Moss, grass, and herbs grow in the first 3 years after the fire, followed by shrubs and saplings. After about 26 years, a secondary forest with dense tree cover has developed. Hardwood and spruce follow about 45–100 years after the fire.

Natural regeneration goes along with continuous LCC, increase in roughness length, and biomass. The LCC subsequent to wildfires alter gradually the heterogeneity of the landscape, vegetation type and fraction, and hence surface properties (e.g., stomatal resistance, roughness length, albedo). The size of the burned areas determines the degree to which surface heterogeneity changes and the atmospheric impacts that will occur as successive landscapes evolve (Mölders and Kramm 2007).

Xue and Shukla (1996) examined the impacts of large-scale sub-Sahara afforestation on climate with the Center for Ocean Land Atmosphere studies (COLA) GCM for June to August. During these months, moisture advection from the Atlantic is the major water vapor source. The African easterly jet at 700–600 hPa governs the circulation and precipitation in this region. Their simulations suggested that afforestation increases precipitation in the afforested area especially in dry years. Precipitation increased more than 20% (0.8 mm/d) in most of the afforested areas expect for the descending branch of the Hadley cell. The afforestation led to reduced precipitation by less than 10% in the region south of the afforestation (Fig. 3.11). The precipitation anomalies

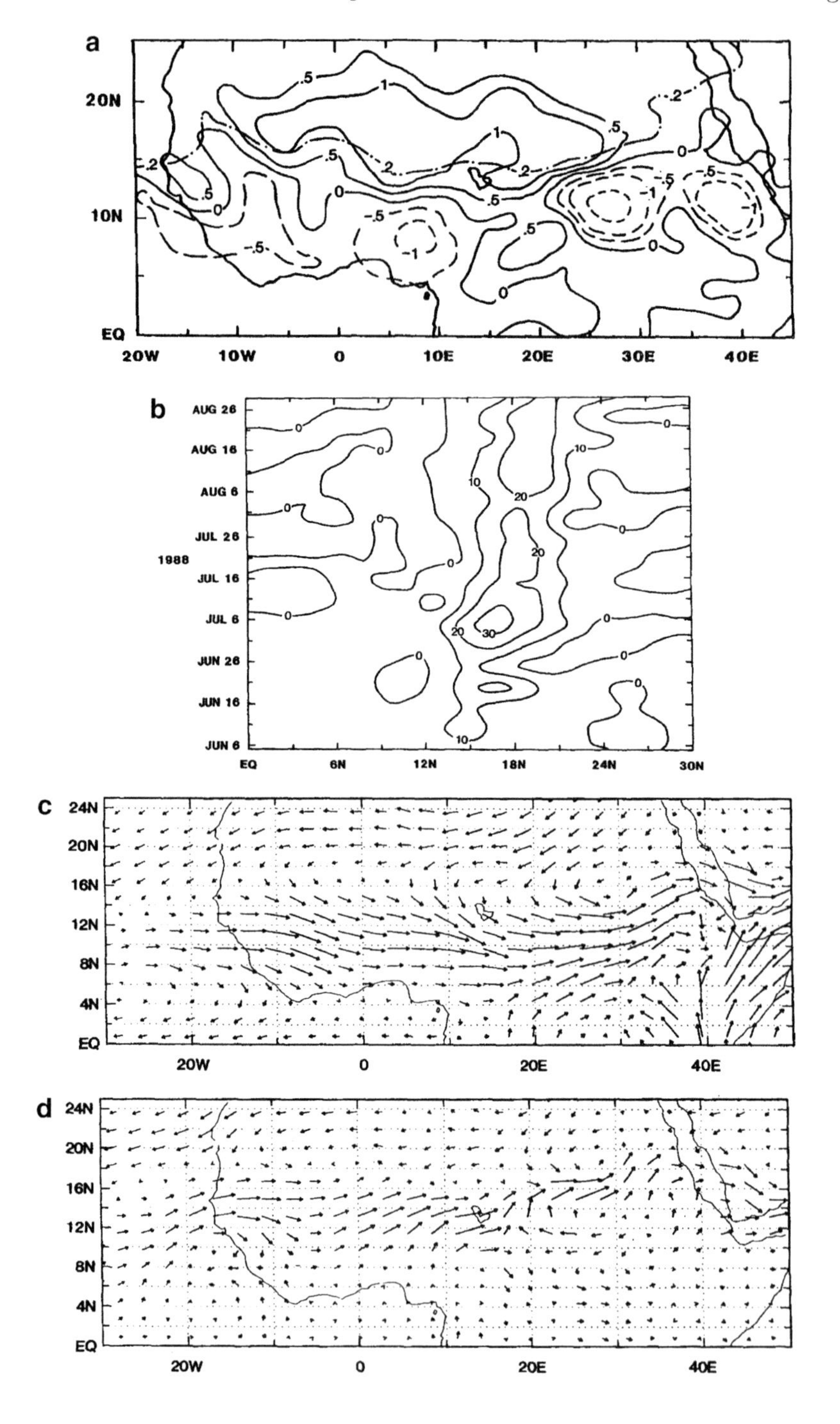
a
20N
10N
EQ
20W
10W
0
10E
20E
30E
40E
b
AUG 26
AUG 16
AUG 6
JUL 26
1988
JUL 16
JUL 6
JUN 26
JUN 16
JUN 6
EQ
6N
12N
18N
24N
30N
c
24N
20N
16N
12N
8N
4N
EQ
20W
0
20E
40E
d
24N
20N
16N
12N
8N
4N
EQ
20W
0
20E
40E

are nonuniform in space and time and mainly due to changes in convection. The altered energy balance modified the regional circulation and subsequently reduced precipitation. The enhanced aerodynamic roughness, and LAI and reduced stomatal resistance led to increased latent heat fluxes. The increased precipitation fed back positively to enhance evapotranspiration further in July and August (Fig. 3.11). The decreased albedo increased absorption of shortwave radiation. The net gain in surface energy resulted in enhanced surface temperatures, upward motions, and convergence.

Since a warmer atmosphere requires more water vapor for saturation to occur than a slightly cooler one, precipitation did not increase in June. In July, however, when evapotranspiration increased appreciably, clouds formed and reduced the temperatures locally by more than 2 K in the lower atmosphere. The moisture convergence became strong enough for enhanced precipitation. Early in July, the cool air below 700 hPa stopped supporting the thermally induced moisture convergence in the coastal area (Fig. 3.11). On the June-July-August (JJA) mean, the moisture convergence was positive in the center of the afforested area and negative south of it (Xue and Shukla 1996).

Afforestation weakens the African easterly jet and strengthens the tropical easterly jet (Fig. 3.11). Observations show that the African easterly jet is strong in dry, and weak in wet years. The opposite is true for the tropical easterly jet. Sensitivity studies showed that the size of the afforested area determines the magnitude of the precipitation changes (Xue and Shukla 1996).

Forestation affects the carbon cycle and other trace gas cycles. Observations in southern Israel, for instance, showed that afforestation of semiarid shrubland reduce biogenic NO emissions from soils (Gelfand et al. 2009). All forestation, especially, growing conifers strongly enhances the VOC concentrations.

The public is most concerned about the link between CO_2 and forest. Since the 1940s, trees have been planted in the USA for carbon sequestration. In midlatitudes, reforestation and fire suppression have shifted the temperate forests to a carbon sink (Bonan 2008). Carbon sequestration is much faster in the tropical than temperate zone due to the more

Figure 3.11. (**a**) JJA mean of precipitation differences afforestation minus control simulations. Contours are at 0, ± 0.5, ± 1, ± 2, ± 4, and ± 8 mm/d. (**b**) Spatiotemporal differences of latent heat fluxes. Contour interval is 10 W m^{-2}. July–August mean of moisture flux (g kg^{-1} m s^{-1}) for (**c**) the control and (**d**) afforestation simulation. Vector length and direction correspond to flux strength and direction, respectively (From Xue and Shukla 1996) favorable climatic conditions. However, the net carbon sequestration is much smaller in the Tropics than midlatitudes as in the Tropics, wood is used primarily as fuel (Bonan 2008).

3.5.4.2 Drylands

Drylands degrade across thresholds into less and less productive lands. Exiting this process needs external intervention (Verstraete et al. 2009). However, such measures require long time before becoming effective. During the intense Sahel drought, for instance, projects were launched in some areas to plant trees to increase precipitation. The long-lasting drought, the growing pressure for food, and the slow growth of trees, however, counteracted this measure as animals grazed the young trees.

The time involved makes it is difficult to assess whether measures to reverse degradations are effective. Long-term (>5 year) monitoring is required due to the high temporal variability of precipitation in drylands. Usually, dryland ecosystems are adapted well to short-term variability. If degradation exceeds a threshold that inhibits the system to recover spontaneously, degradation will continue (Verstraete et al. 2009). The abrupt increase of rainfall in the Sahel in 1994 may be explained by the enhanced warming of the land compared to the ocean that leads to an increased monsoonal landward flow and rainfall (Giannini et al. 2008).

An example of how afforestation, increased cultivation, and restrictions on grazing led to reversal of desertification is the Negev (Fig. 3.1). Here, October rains steeply and the rains of the rest of the rainy season appreciably increased since 1975 as compared to the 20 years before (Otterman et al. 1990). The area that receives 50 mm, the minimum precipitation needed for sufficient plant-available water, has increased. The increased vegetation cover intensified the dynamical convection processes and advection. Inversions capping the ABL penetrated higher due to the enhanced daytime convection. The strengthened advection provided moist air from the warm Mediterranean Sea and contributed to increased precipitation (Otterman et al. 1990).

3.5.5 Urbanization

Urbanization is an extreme case of LCC. Since 1950, the population living in cities has increased by a factor ten. In 2004, 1.2% of the Earth was urban land. Despite urban population grew strongest in the Tropics, most studies on urban effects have been performed for midlatitudes (Fig. 3.12).

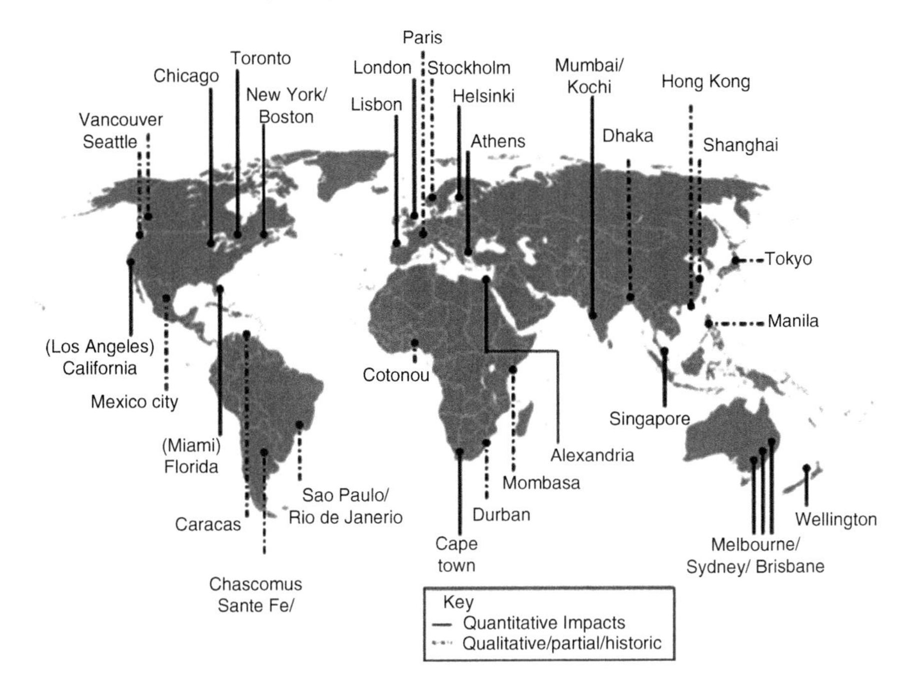

Figure 3.12. Geographical location of city for which major urban studies where performed (From Hunt and Watkiss 2011)

3.5.5.1 Urban Heat Island Effect

The main effects of cities on the airflow are through the drag from buildings due to pressure difference across roughness elements and differential heating of urbanized surfaces, which produces the urban heat island (UHI) effect. To understand the origins of the UHI, we have to consider the surface-heat budget equation. Primary mechanisms to create the UHI effect are the replacement of natural land surfaces by artificial surfaces of different thermal properties (e.g., heat capacity, thermal diffusivity). The urban surfaces heat more intensely than natural surfaces during the day, store the solar energy, and convert it to sensible heat. As sensible heat is transferred to the air, air temperature tends to be 2–10 K higher in urban areas than in adjacent rural areas with consequences for stability. Anthropogenic heat release from emissions adds to the UHI.

The albedo of urban land alters the reflection of incoming solar radiation. In midlatitudes, the albedo of roofs, buildings, and paved streets is typically lower than that of vegetation or bare soil. Consequently, the net energy increases as less solar energy is reflected. The sealing reduces evapotranspiration over the city as compared to an unsealed area with same atmospheric conditions. Evapotranpiration consumes less of the incoming solar energy, and evaporative cooling is low. After precipitation events in a city, most of the water goes into runoff rather than evapotranspiration.

The formation of an UHI effect, its magnitude, and related atmospheric responses depend on the large-scale forcing. The intensity of the UHI is given by the difference between the mean urban and mean rural temperature. It is typically proportional to the city size (Oke 1995). The intensity varies spatially over the city depending on the degree of surface sealing, ventilation, distance to the rural land, and climate region. In arid urban areas, the UHI effect is relatively weak. Here, a city is sometimes rather a heat sink because evapotranspiration from irrigated areas converts huge amounts of solar energy into latent rather than sensible heat. In the subarctic, UHI effects exist, but albedo plays a different role than in temperate or tropical regions.

Whether urbanization enhances the sensible or latent heat fluxes depends on where it occurs. In humid midlatitudes, urbanization shifts the Bowen ratio toward higher values. Here, urbanization reduces evapotranspiration and daytime relative humidity in the city and enhances the UHI effect. The slightly warmer air due to the UHI effect, however, may enhance evapotranspiration in the downwind of the city. Enhanced evapotranspiration and relative humidity were found for cities with water meadows, grassland, lakes, and/or lakes in their downwind region. In semiarid areas, the irrigation of the urban forest and lawns may lead to a shift toward lower Bowen ratios as compared to the adjacent non-irrigated land. Anthropogenic moisture releases from urban irrigation and/or as a by-product of combustion are important sources of enthalpy that result in cooling (Coleman et al. 2010).

The UHI intensity exhibits seasonal cycles and is modulated by cloudiness and wind conditions. The UHI effect has a distinct diurnal cycle with peaks in the late evening to early morning (Fig. 3.13). The positive heat anomaly is most evident during clear, calm nights. Obviously, the UHI intensity differs between weekdays and weekends/holidays. Fujibe (2010) investigated 29 years of hourly data from the Automated Meteorological Data Acquisition System network of Japan and found that on weekends/holidays, the UHI effect was about 0.2–0.25 K and 0.1–0.2 K lower at Tokyo and Osaka, respectively, and about 0.02 K lower at stations located in urban areas with population density of $300–1,000\,\mathrm{km}^{-2}$. On the long term, the UHI effect relatively decreased and increased on Mondays and Fridays, respectively, by about 0.05–

0.1 K/decade in Tokyo and about 0.02 K/decade at stations with population density of 100–1,000 km^{-2}. No significant trends in the differences of the UHI effect existed between weekdays and weekends.

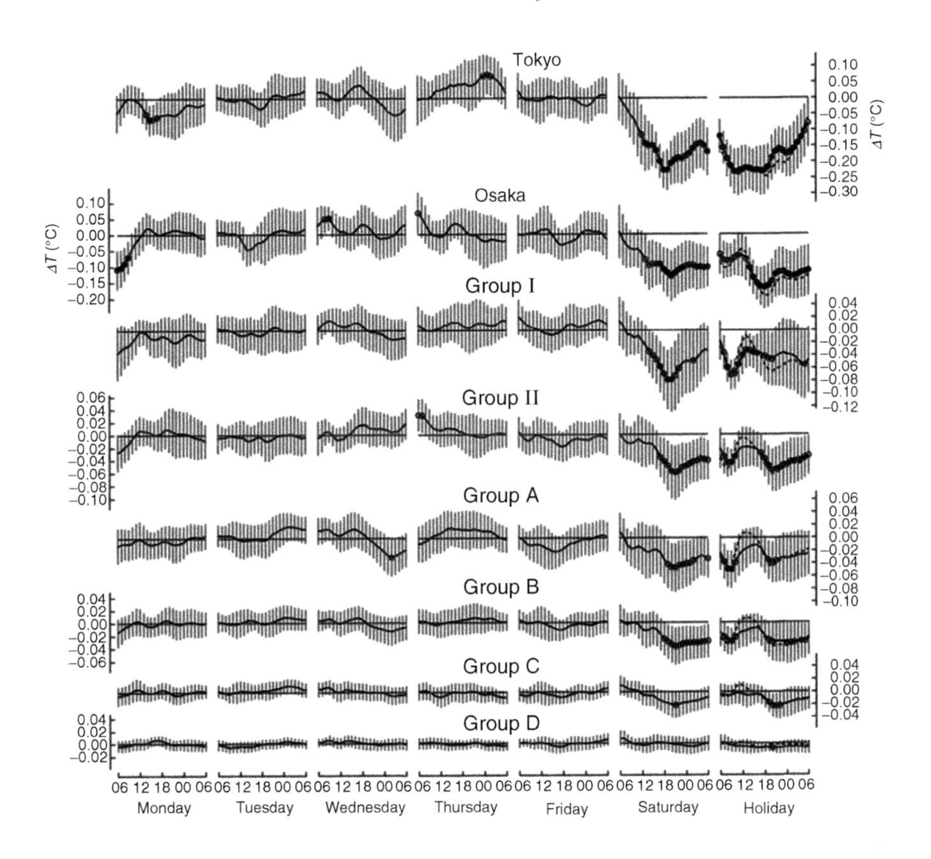

Figure 3.13. Average diurnal variation of the UHI (annual average, in K) at stations of various population density on each day of the week. Open and closed circles indicate UHI effects being significant at the 5% and 1% levels, respectively. In the holiday panels, the dashed lines denote the results for Sundays only. Groups A–D refer to stations with population density ⩾3,000 km^{-2}, 1,000–3,000 km^{-2}, 100–3000 km^{-2}, and 30–100 km^{-2}, respectively. The reference temperature to determine the magnitude of the UHI effect is from stations with population density <100 km^{-2}. Plots base on data collected from March 1979 to February 2008 (From Fujibe 2010)

3.5.5.2 Urban Impacts on Clouds and Precipitations

The pre- and METROMEX investigations suggested that cities may trigger or enhance the formation of convective clouds over areas with enhanced evapotranspiration under favorable large-scale conditions. After

METROMEX, various studies extended the knowledge on urban impacts on precipitation. Changnon (1980), for instance, found that in the downwind of Chicago, summer precipitation increased about 0.4 mm on average from 1931–1976 due to urban growth. Observations show similar precipitation increases for other midlatitude cities (Shepherd 2005): The frequency of late-afternoon storms increased in Phoenix, Arizona, in response to the explosive population growth. Historical records for Mexico City show increased frequency and intensity of showers over the decades as the city grew. Studies on impacts of high-latitude cities on precipitation are still rare. Analyzing precipitation data from 1950 to 2002 for urban and rural sites around Fairbanks, Alaska, Mölders and Olson (2004) found an increase in precipitation due to the growth of Fairbanks. All studies have in common that urbanization increases precipitation amounts in the lee of large conurbations, and it reduces precipitation on the regional average due to the reduced regional-average evapotranspiration.

Urban areas affect precipitation variability by various mechanisms: Advection of the urban heat plume enhances the temperatures downwind of the city. The UHI-thermal perturbation destabilizes the ABL over and in the downwind of the city and promotes vertical lifting (Fig. 3.14). The increased aerodynamic roughness and surface heterogeneity of the city modify the wind field, and lead to low-level convergence with subsequent lifting. Advection of air from the adjacent rural areas replaces the ascending air. Over the city toward the downwind side, the vertical wind component can increase significantly. The water-vapor release associated with combustion humidifies the urban atmosphere. In humid midlatitudes, evapotranspiration over rural land exceeds that over the city leading to a moisture convergence. The large-scale flow and upward lifting of moist, relatively warmer air transport moist air toward the lee side of and into the lee of the city.

If the atmospheric conditions permit saturation to occur, the strong upward transport of moist air can increase the amounts of cloud particles and hydrometeors as compared to the clouds in the undisturbed landscape. Consequently, precipitation may form and/or increase in amount, intensity, and distribution (Changnon and Huff 1986). Over the lifetime of the clouds, more cloud and precipitating particles are formed over and in the lee of the cities than in rural areas. The enhanced moisture convergence and upward transport of water vapor over the city permit the air to reach higher, relatively colder levels. At subfreezing temperatures, this shifts precipitation formation in favor of the more efficient cold path involving riming and the Bergeron–Findeisen process.

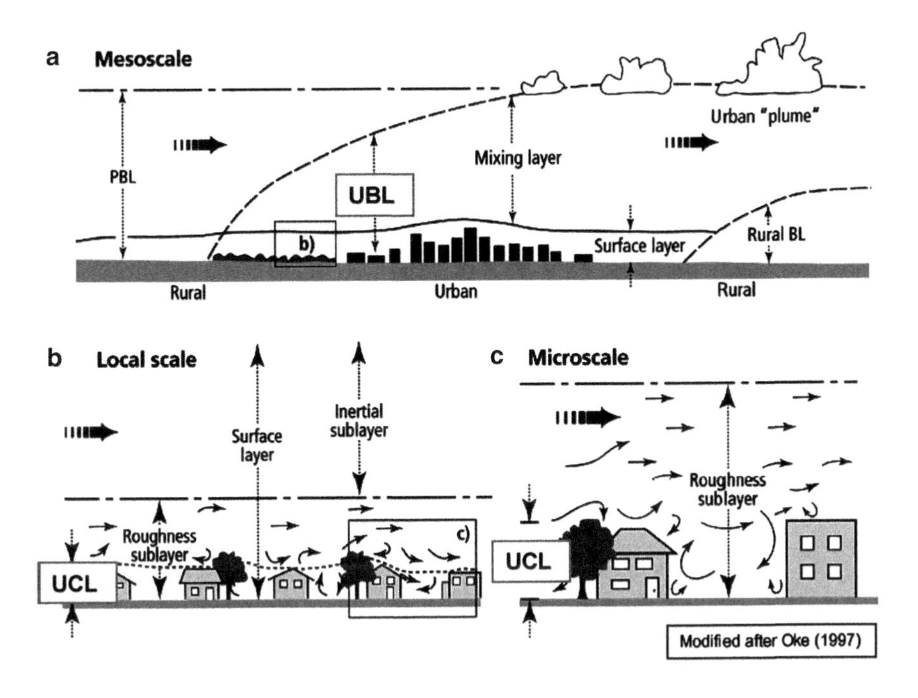

Figure 3.14. Schematic view of modifications of the lower atmosphere in an urban environment as it occurs at various scales (Modified after Oke 1995; from Shepherd 2005)

For a city to affect significantly the precipitation distribution in its major downwind, it must exceed a critical size. In midlatitudes, for instance, urban growth of about 20% may significantly enhance cloud and precipitation formation in the lee of cities of about 300 km^2 in original extent (e.g., Fig. 3.15). The same growth of small settlements hardly affects the atmosphere, except where small settlements grow in an area dominated by grassland (Mölders 1999a). In grassland, evapotranspiration enhances appreciably in the downwind of the grown settlement during the warm seasons.

Anthropogenic heat and water-vapor release from individual point sources, such as power plants, marginally affect precipitation in the immediate vicinity of the stacks due to altered buoyancy. While the impact of individual stacks is negligible most of the time in mid-latitudes, the sum of the anthropogenic heat release within an urban area can contribute substantially to the UHI effect at high latitudes. Anthropogenic water release from combustion may notably affect atmospheric humidity and enhance precipitation in a city's downwind (e.g., Mölders and Olson 2004).

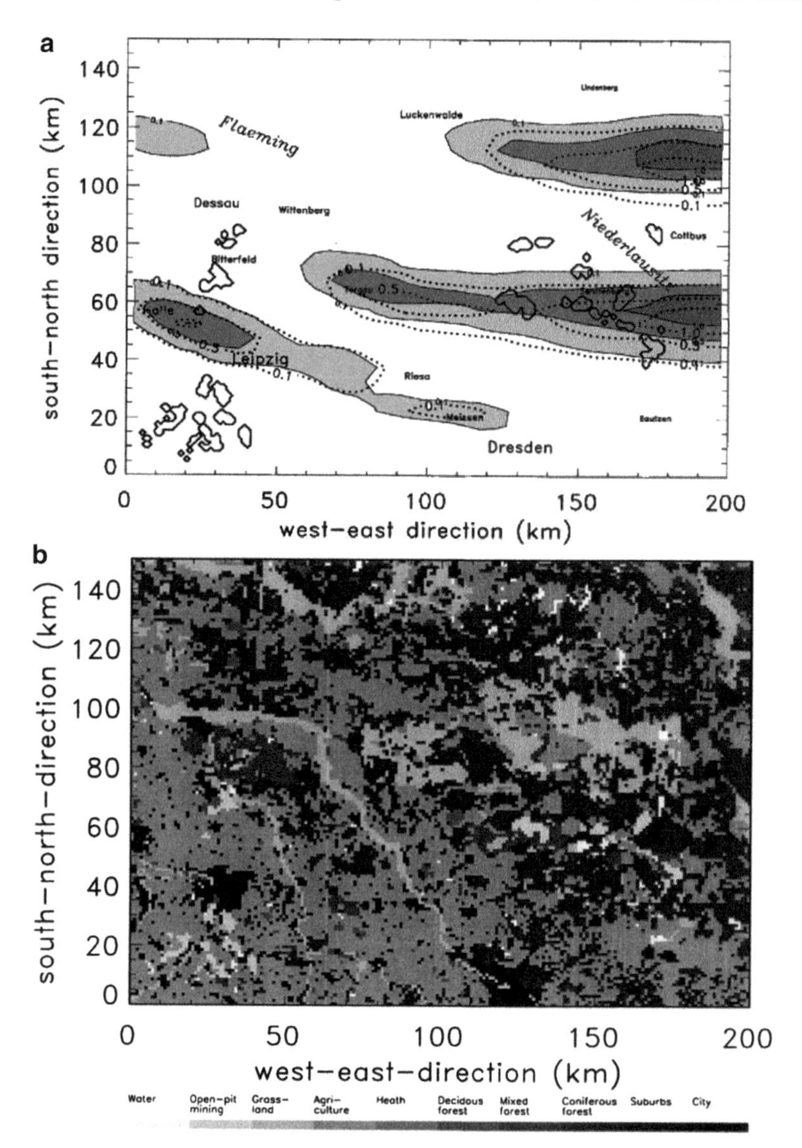

Figure 3.15. (**a**) Comparison of the 24-h accumulated precipitation (mm) as obtained for the reference landscape (**b**) and the landscape with a 20% increase in urban areas in Saxony, Germany, for a convective day in spring, assuming a geostrophic wind from 100° with 7.5 m/s at 3 km height. The names indicate locations of cities, the italic names of regions, and the solid black contours indicate open-pit mines (From Mölders 1999a)

3.5.5.3 Other Dynamical Impacts of Urbanization

The horizontal temperature gradients between urban and rural land can yield NCMC. The partial gradients in equilibrium temperature in combination with the overlying thermodynamic and moisture stratification dominate the upward or downward heat flux of thermally forced systems. Consequently, horizontal temperature gradients evolve that drive the NCMC. The urban circulation depends on the stability of the ABL, and the increase in sensible heat fluxes and roughness as compared to the rural land. Model simulations show that the UHI intensity and ABL stability govern the development of NCMC. Urban-induced NCMC are commonly referred to as UHI circulations. The UHI circulation is stronger during the day than at night because of the stronger urban–rural pressure gradient and vertical mixing during daytime hours. Urban-forced convection is not simply a night–early morning phenomenon.

Observations show that conurbations may delay frontal passages and/or bifurcate or divert precipitating systems (Loose and Bornstein 1977). The UHI effects may interact with and modify classical mesoscale circulations like land/sea breezes (Shepherd 2005) or mountain-valley breezes (Miao et al. 2009). Modifications of the local circulation systems have been modeled and observed, for instance, for Athens, Barcelona, Beijing, Chicago, Houston, Los Angeles, New York, and Paris, just to mention a few.

3.5.5.4 Trace Gas and Aerosol Release from Cities

Urban areas affect the atmosphere also by releasing moisture, trace gases, aerosols, and heat to the atmosphere. Traffic, industrial, and domestic fuel consumption are the major cause for urban air pollution. Thus, urbanization increases the urban emissions over a larger area. As cities grow, CO, NO_x, and particulate matter (PM) emissions, for instance, strongly increase with the increasing number of vehicles.

Airborne pollutants from large cities affect air quality, weather, and climate on the local to continental scale. Locally, the polluted air reduces incoming shortwave radiation with consequences for the energy budget. Urban-induced attenuation is typically less than 10%, but may reach up to 22% and 33% in much polluted megacities like Mexico City or Hong Kong (Kanda 2007).

Over the city and in its lee, the primary pollutants react and form secondary pollutants. The polluted urban air plume affects the natural and agricultural ecosystems in the city's downwind. As conurbations grow, advection of polluted urban air into rural areas, and the impacted downwind region increase. In cities, ABL height and the emissions govern the diurnal cycle of pollutant and PM concentrations. The enhanced vertical mixing and low-level convergence over cities permit pollutants to reach higher levels. Here, the higher wind speeds permit long-range transport of pollutants.

Increased urban emissions increase the availability of CCN and IN from gas-to-particle conversion and direct emissions. Thus, over urban areas and in their polluted downwind, the number of CCNs and INs typically exceeds those in unpolluted rural areas. The increased number of CCNs and INs may counteract the thermodynamic urban effects discussed above and diminish precipitation. Under saturated conditions, the additional CCNs compete for the available excess water vapor. More, but smaller cloud droplets form in the urban air than in the rural environment. Small cloud droplets need longer to form precipitation than large droplets. The size and composition of CCNs influences appreciably the distribution spectrum of cloud droplets. Increases in CCN amounts from pollution have been found to suppress precipitation in winter orographic clouds (Shepherd 2005) and precipitation in general (Cotton and Pielke 2007). In the lee of cities with industries producing large wettable and/or soluble, saturation-pressure reducing aerosols, cloud formation is enhanced frequently. In humid urban air with high SO_2-concentrations, SO_2-molecules serve as CCNs and lead to fog or cloud droplets that reduce visibility. At subzero temperatures, the additional INs over urban areas may affect the partitioning between the ice and liquid phase. If more ice crystals form, the efficiency of precipitation formation increases and may lead to enhanced precipitation downwind of the city. How enhanced CCN and UHI impacts interact with each other is a hot research topic.

3.5.6 Water Bodies

Flooding can have natural and anthropogenic causes. Flooding occurs in response to extreme precipitation events, breakup, coastal storms, or rise of groundwater. Bad water or flood management, straightening of rivers, elimination of flood lands, and insufficient coastal or dike protection enhance the frequency of floods. During high waters of rivers, adjacent marshlands, wetlands, and grasslands can serve as flood land to mitigate medium-size floods. Flooding of large flood-protection areas can avoid flooding in areas that are more critical. Flood prevention by impoundments reduces the peak runoff and increases evaporation on the long term. Planned long-term flooding includes flooding of open-pit mines and creation of water reservoirs for recreational, water management, and/or power generation purposes.

Any flooding replaces the original land-cover with water. The characteristics of land surfaces differ appreciably from those of water (Table 3.1). The changed hydrologic, thermal, and dynamical conditions associated with water may affect notably the stability, water supply to the atmosphere, and cloud and precipitation formation (Fig. 3.16). The magnitude and the kind of atmospheric response to flooding depend on the duration of the flooding, size of the flooded area, the kind of land-surface that was flooded, and the season. In midlatitude summer, cloudiness, for instance, may increase downwind of large flooded areas due to the modification of the moisture convergence (Fig. 3.16), while it may decrease downwind of small, flooded areas due to stabilization (Mölders 1999b).

In the case of short-term flooding events, the atmospheric response depends on the season in which and where the flooding occurs. In midlatitude spring, for instance, when the water is still cooler than the flooded land, the flooded areas evaporate less than the area would have otherwise, while later in fall when the water is warmer than the land, they evaporate more.

Short-term changes in atmospheric conditions due to seasonal flooding do not affect the climate over 30 years, as they are part of its natural variability. The creation of a water reservoir or flooding of an open-pit mine, however, causes persistent LCC. Changes in state variables and fluxes may be visible on the annual, seasonal, monthly, and/or daily average when comparing the 30-year climate averages prior to and after the permanent flooding.

The flooding of appreciably large areas reduces the amplitude of the diurnal air-temperature cycle and daily mean temperatures over and in the downwind of the water surface (Mölders 1999a; Stivari et al. 2005). If the LCC persist, these differences typically do not cancel out and lead to a shift in the local mean temperatures and precipitation.

The impact of the Itaipu Lake formation (1982–1984) is an example for such climate impacts. Stivari et al. (2005) analyzed the impact of this lake formation on temperature and precipitation (Fig. 3.17). They area-averaged the annual values of air temperature (daily average, daily maximum, daily minimum values) at the Itaipu Power Plant, Palotina, and Cascavel that are 2, 35 and 75 km from the lake, respectively, and compared the temperature-deviation amplitude before (1978–1982) and

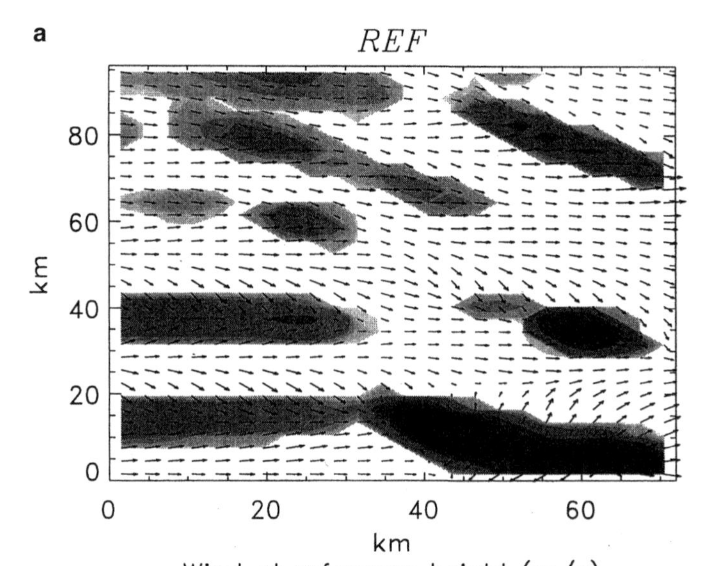

0.001 0.005 0.01 0.05 0.1 0.5 1 1.5

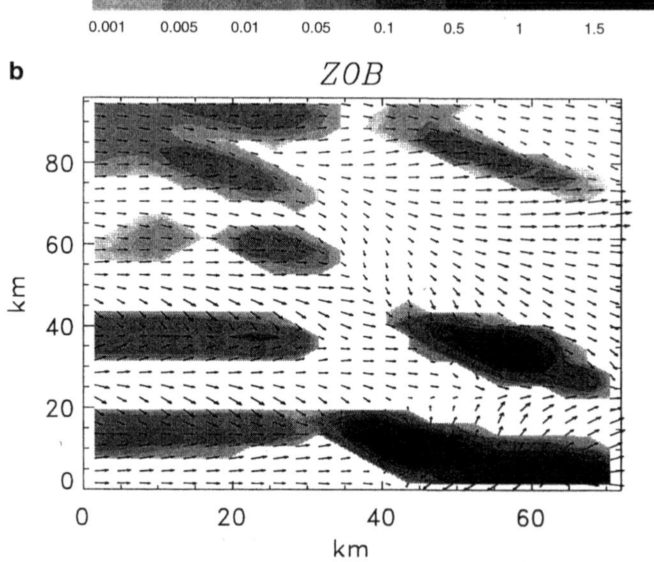

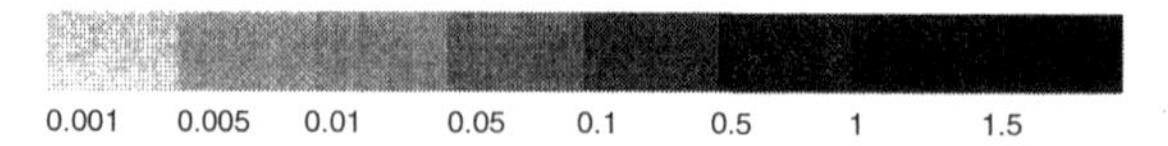

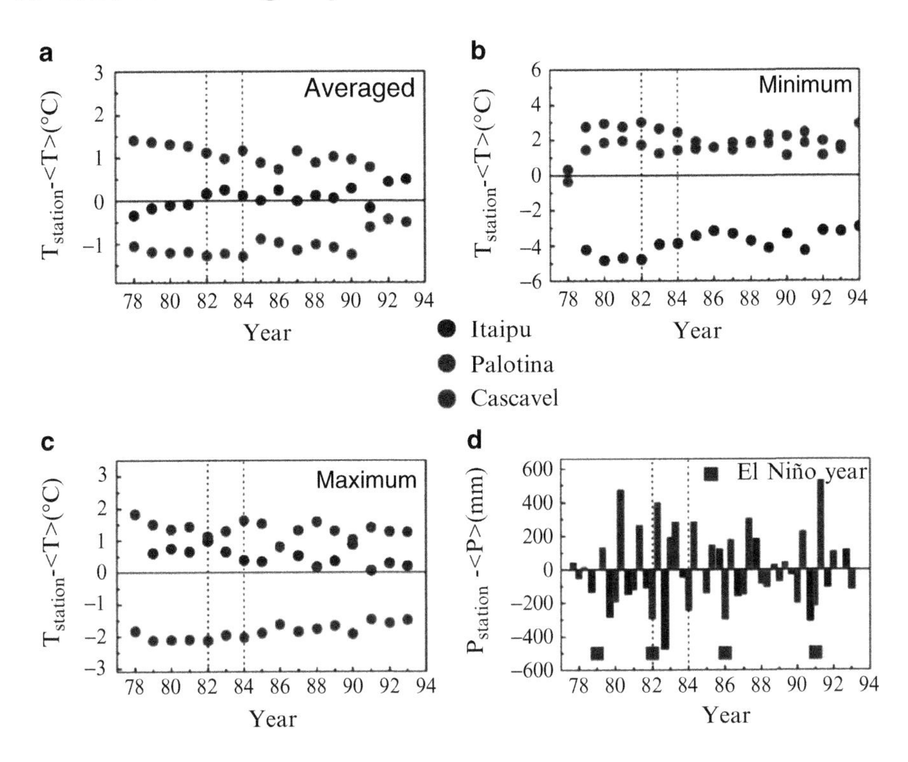

Figure 3.17. Temporal evolution of the deviation between the monthly averaged air temperature (T_{station}) and annual accumulated precipitation at a certain station (P_{station}) and the area-averaged air temperature ($<T>$) and precipitation ($<P>$) over the data of three meteorological stations, Itaipu, Palotina, and Cascavel. (**a**) Average air temperature, (**b**) minimum and (**c**) maximum air temperature, and (**d**) annual accumulated precipitation. The *boxes* indicate El Niño events. The *vertical dotted lines* mark the lake-formation period (Modified after Stivari et al. 2005)

after the lake formation. They found a tendency of a decreased amplitude after the lake formation at all the stations. During the lake formation, a positive (negative) trend deviation from the area average for the daily temperature minimum (maximum) exists at the Itaipu Power Plant, while the opposite is true for the stations farther away from the lake.

Figure 3.16. Vertical sum of cloud-water and ice as obtained for the (**a**) reference landscape and (**b**) landscape with a flooded area of about 20 km width extended from NW to SE between 90 and 50 km in S-N and 15 and 45 km in W–E direction. The smooth water surface reduces wind speed and enhances moisture convergence and cloud formation southeast of the flooded area (From Mölders 1999b)

Such climate changes manifest as follows: Due to the about twice as high heat capacity of water than land, large water bodies store and release heat in the warm and cold seasons, respectively. In the Tropics, large artificial water reservoirs are systematically colder than the land around during daytime, while the opposite is true at night (Stivari et al. 2005). In high and midlatitude summer, the air adjacent to large water reservoirs heats up less as it did without the presence of the water reservoir. The climate becomes more moderate. In mid-latitude winter, the slow release of the stored energy warms the air along the water reservoir's shores. Thus, the land adjacent to the water reservoir experiences milder winters than before. In high latitudes, water reservoirs freeze and are snow-covered in winter. The relatively higher albedo of the snow-covered water reservoir reduces near-surface air temperature via the temperature-albedo feedback. In the case of water reservoirs created by build up of a river, a cold air pool may form over time over the snow-covered water reservoir. The cold air may drain down the river valley and increase the frequency of inversions downstream of the reservoir. The relatively smooth surface of the frozen water reservoir as compared to former boreal forest enhances wind speed locally. In summer, water reservoirs in high latitude may enhance the water supply to the atmosphere.

No matter what caused the flooding, the altered atmospheric water supply can affect cloud and precipitation formation locally. Atmospheric transport can yield significant changes in cloudiness and precipitation even far away from the flooded areas as compared to the landscape prior to the flooding. In semiarid regions, for instance, the construction of lakes and ponds enhances mean dew-point temperatures by 1–2 K under favorable large-scale conditions. The resulting advection of moisture may increase the frequency of fog in their downwind (Stivari et al. 2005).

If artificial water reservoirs are large enough, they may induce classical mesoscale circulations. In 1982, for instance, the Itaipu Lake was created as an artificial water reservoir of 1,460 km^2 (approximately 170 km by 7.5 km) in the Parana River Valley for the Itaipu Power Plant at the Brazilian–Paraguayan border. Analysis of the surface meteorological data (air temperatures, relative humidity, precipitation) and radiosonde soundings available in the area indicate that the Itaipu Lake induces lake breezes (Stivari et al. 2005). Despite lake-breezes are associated with cloud formation, the observations showed no systematic effect of the lake formation on precipitation. Obviously, other local impacts like valley-mountain circulation and El Niño events can mask the impact of lake formation in areas with rain deficit (Stivari et al. 2005).

3.5.7 Interaction of Land-Cover Change Impacts

Most LCC studies examined the impacts of land conversion, that is a "single land-cover change." Herein two different cases were considered: a A converts to a land-cover type Z or several land-cover types A, B, C convert to Z. Such LCC are called *simple* LCC.

In reality, LCC proceed continuously with several types of changes occurring concurrently, that is the LCC are land-cover modifications. *Concurrent* LCC denote changes for which several different land-cover types A, B, C, etc., are converted simultaneously to various land-cover types X, Y, Z, etc. (Mölders 2000b). Such LCC occurred, for instance, in association with the settling of America.

In the light of the extreme nonlinearity of atmospheric processes, it is conceivable that the atmospheric response of concurrent LCC may differ from that of the sum of the response to the simple LCC. Instead, the atmospheric response to the concurrent LCC may diminish or enhance.

The diminution or enhancement of the response to concurrent LCC is difficult to find by comparing observational data before and after the LCC. Consequently, models have been applied to examine the interaction of responses to concurrent LCC. Typically, the impact of the individual LCC is isolated by comparing results of simulations without, with individual LCC, combinations of concurrent LCC to those with all concurrent LCC.

In a linear response, the principle of superposition would apply. At point j, the sum of the differences, $\sum_{i=1}^{n}\left(\chi_j^k - \chi_j^i\right)$, in the simulated state variable or fluxes, χ_i, caused by simple LCC, $i = 1, \ldots n$, would equal to the atmospheric response to the concurrent LCC. This response is given by the difference between the state variables or fluxes in the unchanged landscape χ_j^k and that with the concurrent LCC χ_j^p. Rearranging leads to (Mölders 2000b)

$$(n-1)\chi_j^k - \sum_{i=1}^{n} \chi_j^i + \chi_j^p = \Delta \begin{cases} > 0 & \text{enhancement} \\ = 0 & \text{superposition} \\ < 0 & \text{diminution} \end{cases} \tag{3.5}$$

Here, i stand for the simulations with one of the various simple LCC that led to the new landscape. The indices k and p indicate the simulation with the landscape prior to the changes and with all the concurrent LCC, respectively. A positive (negative) deviation from the principle of superposition indicates enhancement (diminution) of the atmospheric response. Usually, only enhancement/diminution with absolute values greater than the margin of error in routine measurements of the quantity,

χ_j, are considered as relevant. Worth mentioning interaction of responses require $|\Delta|$ to be greater than 0.2 K for air temperature, 0.5 m/s for wind speed, 0.5 g/kg for specific humidity, 0.5 K for soil temperature, and 35 W/m^2 for fluxes (Mölders 2000b).

Typically, deviations from superposition of sensible heat flux, ground-heat flux, net radiation, and surface temperature occur where cloudiness differs strongly between the landscape without and with all concurrent LCC. This means that deviation from superposition often partly result from secondary effects. The size of patches with homogeneous land cover after the LCC must exceed a certain threshold for responses to concurrent LCC to be nonlinear.

Mölders (2000a), for instance, used a mesoscale model to examine the potential atmospheric impacts in spring of the LCC that occurred south of Berlin, Germany, between the 1930s and 1980s. Increase in population, and political and economic changes had altered the landscape dramatically. Marshland was drained to enhance agricultural production. After WWII, urbanization, recultivation of military areas, modified agricultural practices, and enhanced open-pit mining occurred. In total, more than 50% of the landscape experienced LCC. Urbanization and open-pit mining grew at the costs of agriculture. Reforestation and urbanization led to an increase in roughness length from the 1930s to the 1980s. At the same time, overall albedo slightly increased from 0.177 to 0.181. The simulations assumed the same meteorological initial conditions but the landscapes of the 1930s and 1980s. The results showed spatial and temporal changes in the state variables and fluxes and local recycling of previous precipitation. The LCC between the 1930s and 1980s led to decreases in water availability and the temporal and spatial precipitation distribution, intensity, and amount (Fig. 3.18). The resulting slight changes in buoyancy affected formation of convection, time and location of clouds, and precipitation with consequences for the partitioning between evapotranspiration, infiltration, and runoff. The higher evapotranspiration and water availability in the landscape of the 1930s allowed for earlier onset of precipitation and higher regional total precipitation.

The sensitivity of the atmosphere to LCC depends on the land-cover dominating the area. In midlatitudes in spring, LCC in regions dominated by forest and grassland cause a stronger atmospheric response than in regions dominated by agriculture. Urbanization in a landscape with open-pit mines or with flooded open-pit mines increases cloudiness in the lee of large cities, but reduces the regional average of daily accumulated precipitation. In the landscape with open-pit mines,

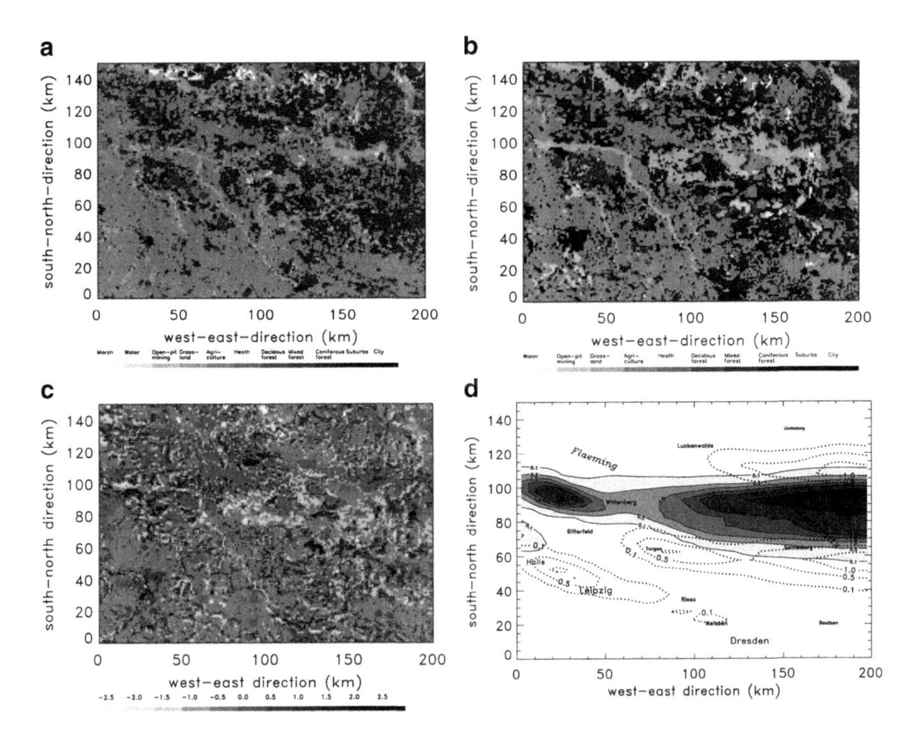

Figure 3.18. Distribution of land-cover for (**a**) the 1930s and (**b**) the 1980s. The area encompasses 30,000 km^2 (51.8009N, 11.8589E; 52.8179N, 11.8539E; 51.8009N, 14.8549E; 52.8179N, 14.8499E) south of Berlin, Germany. (**c**) Differences in 24-h accumulated evapotranspiration (1930–1980) and (**d**) 24-h accumulated precipitation as obtained by the simulation assuming the landscape of the 1930s (*gray shaded*) and the 1980s (*dotted lines*). Maximum values are 4.3 and 1.4 mm for the simulation with the landscape of the 1930s and 1980s, respectively (From Mölders 2000a)

urbanization leads to precipitation over and downwind of the grown city, while no clouds and precipitation occur in the lee of the city without open-pit mines (Mölders 1999a).

When concurrent LCC occur on a continental-scale, the changes in local climate differ among regions and quantities. Since the settlement of North America, a large fraction of the contiguous USA was converted for agricultural production. Copeland et al. (1996) performed simulations with the Regional Atmospheric Modeling System (RAMS) alternatively assuming the modern land-cover and that before the settling of America. To avoid differences due to climate changes, they run both simulations for July 1989. The continental-wide LCC associated with the settling of America caused coherent regions of substantial changes in 2-m temperature, humidity, wind, and precipitation. The changes are of both negative and positive sign and can be related to changes in vegetation parameters such as albedo, roughness length, LAI, and vegetation fraction.

Copeland et al. (1996) found that changes in mixing ratios reached up to 1g/kg or 10% locally. In most regions, the increased humidity results mainly from enhanced evapotranspriation due to temperature increases. In some regions, where cropland replaced the former short grass or steppe, the increased evapotranspiration relates to the increase in LAI. Decreases in wind speeds occurred mainly where the natural short-grass prairie or semidesert changed to shrub and cropland, but these decreases remained lower than 10%. The greatest increases in wind speed (up to 46%) occurred where cropland replaced forest as in the Great Lakes region, Florida, and the southeastern coastal regions. Daily precipitation increased the strongest in Florida (0.96 mm/d; 126%), followed by California (0.06 mm/d; 55%), and New England (0.16 mm/d; 39%). In about half of the remaining regions, precipitation increased about 5–10% as compared to values obtained for the natural landscape. The precipitation changes are related mainly to changes in convection. Convection changed due to differences in stability and low-level moisture convergence resulting from the altered near-surface temperature, humidity, and wind. Due to the LCC, the near-surface atmosphere over the contiguous USA became slightly moister (0.26 g/kg; 2%).

The general tenor of studies on concurrent land-cover changes are consistent with similar studies performed with other models (e.g., Diffenbaugh 2009). However, the quantitative changes differ due to other parameterizations and parameters used, differences in land-cover data, episodes simulated, length of simulation, and model resolution. The precipitation changes reported for Florida by Copeland et al. (1996), for instance, exceed those obtained with RAMS when assuming the 1900, 1973, and 1993 landscapes of southern Florida (Pielke et al. 1999). The latter study found decreases in simulated average summer rainfall of 9% and 11% for the simulations with the 1973 and 1993 landscapes as compared with the simulation with the 1900 landscape. Major reasons for the discrepancies are the different model resolution and the more drastic LCC assumed by Copeland et al. (1996) than assumed by Pielke et al. (1999).

References

Amiro BD, MacPherson JI, Desjardins RL (1999) BOREAS flight measurements of forest-fire effects on carbon dioxide and energy fluxes. Agric For Meteorol 96:199–208

André J-C, Bougeault P, Goutorbe J-P (1990) Regional estimates of heat and evaporation fluxes over non-homogeneous terrain. Examples from the HAPEX-MOBILHY programme. Bound Layer Meteorol 50:77–108

Anthes RA (1984) Enhancement of convective precipitation by mesoscale variations in vegetative covering in semiarid regions. J Clim Appl Meteorol 23:541–554

Avissar R (1993) Observations of leaf stomatal conductance at the canopy scale: an atmospheric modeling perspective. Bound Layer Meteorol 64:127–148

Avissar R, Werth D (2005) Global hydroclimatological teleconnections resulting from tropical deforestation. J Hydromet 6:134–145

Baidya Roy S, Avissar R (2002) Impact of land use/land cover change on regional hydrometeorology in Amazonia. J Geophys Res 107. doi:8037 10.1029/2000jd000266

Bolle H-J (1995) Identification and observation of desertification processes with the aid of measurements from space: results from the European Field Experiment in Desertification-threatened Areas (EFEDA). Environ Monit Assess 37:93–101

Bonan GB (2008) Forests and climate change: forcings, feedbacks, and the climate benefits of forests. Science 320:1444–1449

Changnon SA (1980) Evidence of urban and lake influences on precipitation in the Chicago area. J Appl Meteorol 10:1137–1159

Changnon SA, Huff FA (1986) The urban-related nocturnal rainfall anomaly at St. Louis. J Clim Appl Meteorol 25:1985–1995

Charney J, Stone PH, Quirk WJ (1975) Drought in the Sahara: a biogeophysical feedback mechanism. Science 187:434–435

Chase TN, Pielke RA Sr, Kittel TGF, Baron JS, Stohlgren TJ (1999) Potential impacts on Colorado Rocky Mountain weather due to land use changes on the adjacent Great Plains. J Geophys Res 104:16673–16690

Chase TN, Pielke Sr RA, Kittel TGF, Nemani RR, Running SW (2000) Simulated impacts of historical land cover changes on global climate in northern winter. Clim Dyn 16:93–105

Chen F, Avissar R (1994) The impact of land-surface wetness heterogeneity on mesoscale heat fluxes. J Appl Meteorol 33:1323–1340

Coleman RF, Drake JF, McAtee MD, Belsma LO (2010) Anthropogenic moisture effects on WRF summertime surface temperature and mixing ratio forecast skill in Southern California. Wea Forecast 25:1522–1535

Collins DC, Avissar R (1994) An evaluation with the Fourier amplitude sensitivity test (FAST) of which land-surface parameters are of greatest importance in atmospheric modeling. J Clim 7:681–703

Copeland JH, Pielke RA, Kittel TGF (1996) Potential climatic impacts of vegetation change: a regional modeling study. J Geophys Res 101:7409–7418

Coppin P, Jonckheere I, Nackaerts K, Muys B (2004) Digital change detection methods in ecosystem monitoring: a review. Int J Remote Sens 25:1565–1596

Corbera E, Estrada M, Brown K (2010) Reducing greenhouse gas emissions from deforestation and forest degradation in developing countries: revisiting the assumptions. Clim Change 100:355–388

Cotton WR, Pielke RA (2007) Human impacts on weather and climate. Cambridge University Press, Cambridge

Crook NA (1996) Sensitivity of moist convection forced by boundary layer processes to low-level thermodynamic fields. Mon Wea Rev 124:1767–1785

D'Almeida C, Vörösmarty CJ, Hurtt GC, Marengo JA, Dingman SL, Keim BD (2007) The effects of deforestation on the hydrological cycle in Amazonia: a review on scale and resolution. Int J Climatol 27:633–647

Diffenbaugh NS (2009) Influence of modern land cover on the climate of the United States. Clim Dyn 33:945–958

Euskirchen ES, McGuire AD, Rupp TS, Chapin FSI, Walsh JE (2009) Projected changes in atmospheric heating due to changes in fire disturbance and the snow season in the western Arctic, 2003–2100. J Geophys Res 114:G04022. doi:10.1029/2009JG001095

Fang S, Mu Y (2007) NO_x fluxes from three kinds of agricultural lands in the Yangtze Delta, China. Atmos Environ 41:4766–4772

Friedrich K, Mölders N (2000) On the influence of surface heterogeneity on latent heat fluxes and stratus properties. Atmos Res 54:59–85

Fujibe F (2010) Day-of-the-week variations of urban temperature and their long-term trends in Japan. Theor Appl Climatol 102:393–401

Gash JHC, Kabat P, Monteny BA, Amadou M, Bessemoulin P, Billing H, Blyth EM, deBruin HAR, Elbers JA, Friborg T, Harrison G, Holwill CJ, Lloyd CR, Lhomme JP, Moncrieff JB, Puech D, Soegaard H, Taupin JD, Tuzet A, Verhoef A (1997) The variability of evaporation during the HAPEX-Sahel intensive observation period. J Hydrol 188–189:385–399

Gelfand I, Feig G, Meixner FX, Yakir D (2009) Afforestation of semi-arid shrubland reduces biogenic NO emission from soil. Soil Biol Biochem 41:1561–1570

Giannini A, Biasutti M, Verstraete MM (2008) A climate model-based review of drought in the Sahel: desertification, the re-greening and climate change. Global Planet Change 64:119–128

Guenther A (1997) Seasonal and spatial variations in natural volatile organic compound emissions. Ecol Appl 7:34–45

Gulden LE, Yang Z-L, Niu G-Y (2008) Sensitivity of biogenic emissions simulated by a land-surface model to land-cover representations. Atmos Environ 42:4185–4197

Hély C, Flannigan M, Bergeron Y, McRae D (2001) Role of vegetation and weather on fire behavior in the Canadian mixed wood boreal forest using two fire behavior prediction systems. Can J For Res 31:430–441

Henderson-Sellers A, Dickinson RE, Durbidge TB, Kennedy P, McGuffie K, Pitman A (1993) Tropical deforestation-modeling local to regional scale climate change. J Geophys Res 98:7289–7315

Hunt A, Watkiss P (2011) Climate change impacts and adaptation in cities: a review of the literature. Clim Change 104:13–49

Huntington TG (2006) Evidence for intensification of the global water cycle: review and synthesis. J Hydrol 319:83–95

Jonko A, Hense A, Feddema J (2010) Effects of land cover change on the tropical circulation in a GCM. Clim Dyn 35:635–649

Kanda M (2007) Progress in urban meteorology: a review. J Meteorol Soc Jpn 85B:363–383

Keppler F, Hamilton JTG, Braß M, Rockmann T (2006) Methane emissions from terrestrial plants under aerobic conditions. Nature 439:187–191

Kesselmeier J, Staudt M (1999) Biogenic Volatile Organic Compounds (VOC): an overview on emission, physiology and ecology. J Atmos Chem 33:23–88

Kim H, Kim S, Dale BE (2009) Biofuels, land use change, and greenhouse gas emissions: some unexplored variables. Environ Sci Technol 43:961–967

Kueppers LM, Snyder MA, Sloan LC, Cayan D, Jin J, Kanamaru H, Kanamitsu M, Miller NL, Tyree M, Du H, Weare B (2008) Seasonal temperature responses to land-use change in the western United States. Glob Planet Change 60:250–264

Lauwaet D, van Lipzig N, Kalthoff N, De Ridder K (2010) Impact of vegetation changes on a mesoscale convective system in West Africa. Meteorol Atmos Phys 107:109–122

Lavender SL, Taylor CM, Matthews AJ (2010) Coupled land–atmosphere intraseasonal variability of the West African Monsoon in a GCM. J Clim 23:5557–5571

LeMone MA, Grossman RL, Coulter RL, Wesley ML, Klazura GE, Poulos GS, Blumen W, Lundquist JK, Cuenca RH, Kelly SF, Brandes EA, Oncley SP, McMillen RT, Hicks BB (2000) Land–atmosphere interaction research, early results, and opportunities in the Walnut River watershed in Southeast Kansas: CASES and ABLE. Bull Am Meteorol Soc 81:757–779

Li Z (2007) Investigations on the impacts of land-cover changes and/or increased CO_2 concentrations on four regional water cycles and their interaction with the global water cycle. Ph.D. thesis, Department of Atmospheric Sciences, University of Alaska Fairbanks, 329 pp

Li Z, Molders N (2008) Interaction of impacts of doubling CO_2 and changing regional land-cover on evaporation, precipitation, and runoff at global and regional scales. Int J Climatol 28:1653–1679

Loose T, Bornstein RD (1977) Observations of mesoscale effects on frontal movement through an urban area. Mon Wea Rev 105:563–571

Miao S, Chen F, LeMone MA, Tewari M, Li Q, Wang Y (2009) An observational and modeling study of characteristics of urban heat island and boundary layer structures in Beijing. J Appl Meteorol Climatol 48:484–501

Mölders N (1999a) On the atmospheric response to urbanization and open-pit mining under various geostrophic wind conditions. Meteorol Atmos Phys 71:205–228

Mölders N (1999b) On the effects of different flooding stages of the Odra and different land-use types on the local distributions of evapotranspiration, cloudiness and rainfall in the Brandenburg-Polish border area. Contrib Atmos Phys 72:1–24

Mölders N (2000a) Similarity of microclimate as simulated over a landscape of the 1930s and the 1980s. J Hydromet 1:330–352

Mölders N (2000b) Application of the principle of superposition to detect nonlinearity in the short-term atmospheric response to concurrent land-use changes associated with future landscapes. Meteorol Atmos Phys 72:47–68

Mölders N (2001) On the uncertainty in mesoscale modeling caused by surface parameters. Meteorol Atmos Phys 76:119–141

Mölders N (2005) Plant and soil parameter caused uncertainty of predicted surface fluxes. Mon Wea Rev 133:3498–3516

Mölders N, Kramm G (2007) Influence of wildfire induced land-cover changes on clouds and precipitation in Interior Alaska – a case study. Atmos Res 84:142–168

Mölders N, Olson MA (2004) Impact of urban effects on precipitation in high latitudes. J Hydromet 5:409–429

Mölders N, Strasser U, Schneider K, Mauser W, Raabe A (1997) A sensitivity study on the initialization of surface characteristics in meso-γ/β-modeling using digitized vs. satellite derived land-use data. Contrib Atmos Phys 70:173–187
Monteny BA, Lhomme JP, Chehbouni A, Troufleaub D, Amadouc M, Sicota M, Verhoefd A, Galle S, Said F, Lloyd CR (1997) The role of the Sahelian biosphere on the water and the CO_2 cycle during the HAPEX-Sahel experiment. J Hydrol 188–189:516–535
Nicholson SE, Tucker CJ, Ba MB (1998) Desertification, drought, and surface vegetation: an example from the West African Sahel. Bull Am Meteorol Soc 79:815–829
Oke TR (1995) Boundary layer climates. Routledge, New York
Otterman J, Manes A, Rubin S, Alpert P, Starr D (1990) An increase of early rains in southern Israel following land-use change? Bound Lay Meteorol 53:333–351
Pederson JR, Massman WJ, Mahrt L, Delany A, Oncley S, Hartog GD, Neumann HH, Mickle RE, Shaw RH, Paw U KT, Grantz DA, Macpherson JI, Desjardins R, Schuepp PH, Pearson R, Arcado TE (1995) California ozone deposition experiment: methods, results, and opportunities. Atmos Environ 29:3115–3132
Pielke RA, Walko RL, Steyaert LT, Vidale PL, Liston GE, Lyons WA, Chase TN (1999) The influence of anthropogenic landscape changes on weather in south Florida. Mon Wea Rev 127:1663–1673
Pielke RA, Adegoke J, Beltrán-Przekurat A, Hiemstra CA, Lin J, Nair US, Niyogi D, Nobis TE (2007) An overview of regional land-use and land-cover impacts on rainfall. Tellus B 59:587–601
Ramos da Silva R, Avissar R (2006) The hydrometeorology of a deforested region of the Amazon basin. J Hydromet 7:1028–1042
Ramos da Silva R, Werth D, Avissar R (2008) Regional impacts of future land-cover changes on the Amazon basin wet-season climate. J Clim 21:1153–1170
Sassen K (2005) Meteorology: dusty ice clouds over Alaska. Nature 434:456–456
Saulo C, Ferreira L, Nogués-Paegle J, Seluchi M, Ruiz J (2010) Land–atmosphere interactions during a northwestern Argentina low event. Mon Wea Rev 138:2481–2498
Schädler G, Kalthoff N, Fiedler F (1990) Validation of a model for heat, mass and momentum exchange over vegetated surfaces using LOTREX-10E/HIBE88 data. Contrib Atmos Phys 63:85–100
Schubert SD, Suarez MJ, Pegion PJ, Koster RD, Bacmeister JT (2004) On the Cause of the 1930s Dust Bowl. Science 303:1855–1859
Segal M, Avissar R, McCumber MC, Pielke RA (1988) Evaluation of vegetation effects on the generation and modification of mesoscale circulations. J Atmos Sci 45:2268–2292
Sellers PJ, Hall FG, Asrar G, Strebel DE, Murphy RE (1992) An overview of the First International Satellite Land Surface Climatology Project (ISLSCP) Field Experiment (FIFE). J Geophys Res 97:18345–18371
Sellers PJ, Hall FG, Kelly RD, Black A, Baldocchi D, Berry J, Ryan M, Ranson KJ, Crill PM, Lettenmaier DP, Margolis H, Cihlar J, Newcomer J, Fitzjarrald D, Jarvis PG, Gower ST, Halliwell D, Williams D, Goodison B, Wickland DE, Guertin FE (1997) BOREAS in 1997: experiment overview, scientific results, and future directions. J Geophys Res 102:28731–28769
Shepherd JM (2005) A review of current investigations of urban-induced rainfall and recommendations for the future. Earth Interact 9:1–27

Shukla J, Nobre C, Sellers P (1990) Amazon deforestation and climate change. Science 247:1322–1325

Shuttleworth WJ (1988) Macrohydrology – the new challenge for process hydrology. J Hydrol 10:31–56

Sivakumar MVK (2007) Interactions between climate and desertification. Agric For Meteorol 142:143–155

Stivari S, Oliveira A, Soares J (2005) On the climate impact of the local circulation in the Itaipu Lake area. Clim Change 72:103–121

Taylor CM, Parker DJ, Harris PP (2007) An observational case study of mesoscale atmospheric circulations induced by soil moisture. Geophys Res Lett 34. doi:L15801 10.1029/2007gl030572

Verstraete MM, Scholes RJ, Smith MS (2009) Climate and desertification: looking at an old problem through new lenses. Front Ecol Environ 7:421–428

Wallace JS, Wright IR, Stewart JB, Holwill CJ (1991) The Sahelian Energy Balance Experiment (SEBEX): ground based measurements and their potential for spatial extrapolation using satellite data. Adv Space Res 11:131–141

Werth D, Avissar R (2002) The local and global effects of Amazon deforestation. J Geophys Res 107(D20): 8087. doi:8010.1029/2001JD000717

Wright IR, Gash JHC, Rocha HRD, Shuttleworth WJ, Nobre CA, Maitelli GT, Zamparoni CAGP, Carvalho PRA (1992) Dry season micrometeorology of central Amazonian ranchland. Q J R Meteorol Soc 118:1083–1099

Xue Y (1996) The impact of desertification in the Mongolian and the inner Mongolian grassland on the regional climate. J Clim 9:2173–2189

Xue Y, Shukla J (1996) The influence of land surface properties on Sahel climate. Part II. Afforestation. J Clim 9:3260–3275

Zhou L, Dickinson RE, Tian Y, Fang J, Li Q, Kaufmann RK, Tucker CJ, Mynen RB (2004) Evidence for a significant urbanization effect on climate in China. Proc Natl Acad Sci USA 101:9540–9954

Chapter 4

Future Challenges

Since 1930, the world population more than tripled. Consequently, energy and food demands have increased, which has resulted in more land-cover changes (LCC) for food production and housing. The percentage of people living in cities increased from about 3% of the world's population in 1800 to 47% in 2000. Since 1950, the population living in cities increased by a factor of 10 meaning that the rate of this increase is highly nonlinear. This development went along with the increasing of the mean size of cities. In 1950, 83 cities had populations greater than one million people, while in 2007, already 468 cities had such population numbers (`http://www.citypopulation.de/world/Agglomerations.html`). With this trend continuing, urban population doubles every 38 years. In 2010, 25 megacities (i.e., metropolitan areas with population greater than ten million) existed worldwide as either a single metropolitan area or the result of converging of two or more metropolitan areas or urban agglomerations. The United Nations estimate that urban population will reach about five billion by 2030. The growth of existing and formation of new megacities goes along with an increase in anthropogenic emissions of trace gases, particles, heat and water vapor due to traffic, domestic heating, industrial and energy production.

Typically, urbanization has economic and/or political drivers. The rapid economic development in many Asian countries in the last decade led to urban growth, LCC and enhanced emissions. Urbanization occurred mostly at the cost of arable land; grassland has been cultivated,

N. Mölders, *Land-Use and Land-Cover Changes*, Atmospheric and Oceanographic Sciences Library 44, DOI 10.1007/978-94-007-1527-1_4,

and in tropical areas, deforestations occurred. These LCC will affect sustainable future development.

Besides urbanization and increased use of land for agriculture, future LCC may occur from indirect forces like an increased population pressure, a changing economy pushing toward market crops, the availability of new technologies or capital, changes in land-ownership, government, or regulations of land. New mechanical equipment and/or fertilizers may become available to cultivate formerly unattractive land.

In addition to anthropogenic LCC, natural LCC may occur with some of them being more frequent due to climate change. Various climate projections anticipate a warming and changes in precipitation in response to increased emissions of green-house gases (GHG). Future LCC and their impacts on the atmosphere will differ among climatic regions, landscape types, land-use practices, and economic and political constraints. Glaciers, for instance, may experience retreat due to warmer climate conditions and/or decrease in precipitation. Mountainous regions may see shorter snow seasons, higher rainfall amounts and/or rates and erosion, and soil degradation with consequences for land-use. The longer or more frequent summer droughts expected in Mediterranean regions, for instance, may lead to water shortage, salinization of irrigated areas, and abandoning of rain-fed or irrigated agriculturally used land. The expected raise of the sea level may claim land in deltas and coastal lowlands.

Natural temporal LCC (e.g., flooding and wildfires) require immediate attention, while large-scale processes (e.g., climate changes) are of societal relevance for planning. Thus, major future challenges associated with LCC are the increasing land demands for food and biofuel production, increasing emissions, and the atmospheric conditions in megacities. Another challenge is the alarming increase of natural fire regimes in the last decades with their threats to native biodiversity and human well-being. Increases in temperatures and length of fire season led to increases in area-burned and fire severity. The expected future increase of fire activity in response to climate warming (Soja et al. 2007) requires preparations for prolonged fire seasons. Fire-suppression policies need to be adapted especially for fires upwind of populated areas.

4.1 Future Emission Scenarios

Eco-socio demands and political decisions drive migration into cities, LCC, and related emissions. The LCC and changed emissions may even have caused those demands and decisions. The future magnitude of migration and LCC, the type of land-cover and related emission changes, and their temporal-spatial evolutions are unknown. These changes have to be estimated based on the best scientific knowledge and reasonable assessment of future eco-social demands and political development and have to be updated on a regular basis.

The Intergovernmental Panel on Climate Change (IPCC) issued a Special Report on Emissions Scenarios (SRES). The SRES replaced the six IS92-scenarios previously used. All these emission scenarios as well as pure academic assumptions on CO_2 increases (e.g., doubling, tripling) have been used widely to investigate potential future impacts of LCC under altered climate conditions.

All IS92-scenarios reflect the large uncertainty associated with the evolution of population and economic growth, advance in technology, technology transfer, and responses to environmental and economic constraints. They differ in their assumptions on population and economy development. The IS92A scenario, often-called business-as-usual scenario, for instance, assumes an annual 1% increase in CO_2 from 1990 resulting in a near doubling of the CO_2 concentration by 2050.

The SRES scenarios differ in their assumptions on trends of population growth, economic development, and concurrent regional and local responses of agriculture to the changing climate. Agricultural production adapts to increasing temperatures and CO_2 levels and changing precipitation regimes. The response of agricultural production to climate change differs regionally, because climate change may affect agricultural production positively in some and negatively in others regions at some or the same time. Assumed socioeconomic factors determine the coping with changes in crop productivity, price, and demand of products.

Each scenario provides a storyline of changes depending on the assumed global and regional population growths, energy demands, and development of technology and economics. These assumptions have consequences for land-cover and biogenic and anthropogenic emissions. The major scenarios each have an underlying theme but address aspects within that theme differently (Fig. 4.1). The A-scenarios focus more on economics and the B-scenarios on the environment. Figure 4.2 illustrates, exemplarily, the CO_2 emissions associated with the A1B-, A2-, and B1-scenarios.

The A1-scenario family assumes rapid economic growth with a quick spread of new and efficient technologies. The global population reaches nine billion in 2050 and then gradually declines. Income and way-of-life converge worldwide. Subsets assume an emphasis on fossil fuels (A1FI), all energy sources (A1B), and non-fossil energy sources.

The A2-scenario family assumes a heterogeneous world with self-reliance, preservation of local identities, a continuously increasing global

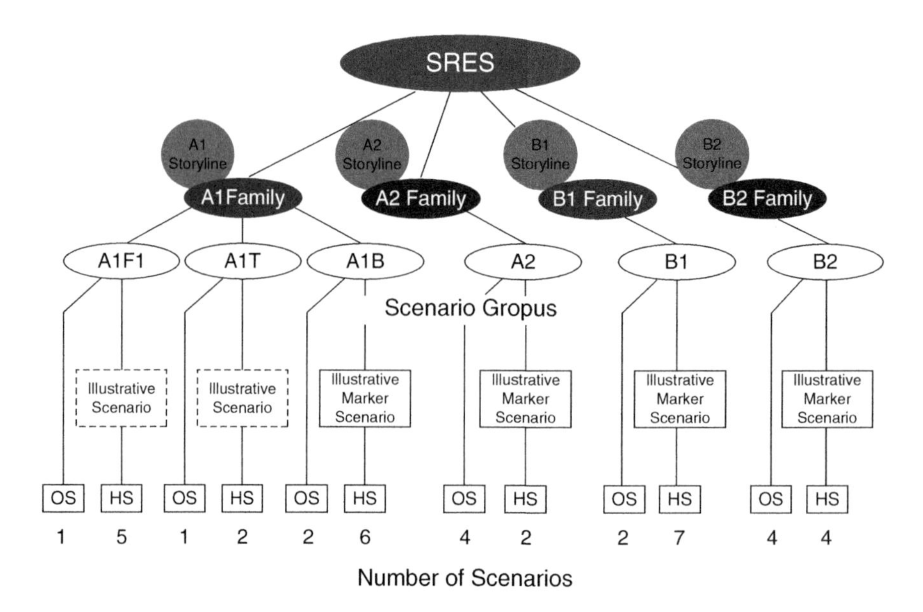

Figure 4.1. Schematic view of the SRES families (Modified after SRES 2000)

population with slow changes in fertility pattern, and primarily regionally oriented economic development and technological change. CO_2 concentrations increase from 380 to 800 ppm between 2000 and 2100.

The B1-scenario family assumes a "globalized" world with global solutions to economic, social, and environmental stability. It assumes a rapid shift toward a service and information economy with resource-efficient and clean technologies. Like in the A1-scenario, the population rises to nine billion in 2050 and then declines.

The B2-scenario family assumes a more "divided" world than the B1-scenarios with emphasis on local solutions to economic, social, and

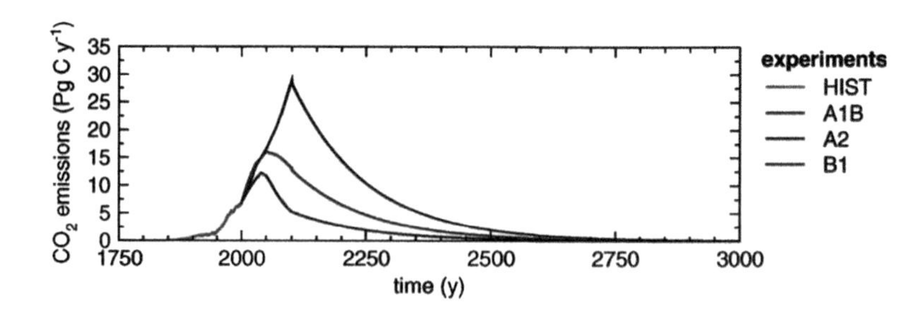

Figure 4.2. CO_2 emissions as prescribed in the Intergovernmental Panel on Climate Change (IPCC) Special Report on Emissions Scenarios (SRES) scenario experiments (From Schurgers et al. 2008)

environmental stability. Like the B1-scenarios, it is economic friendly, but develops less rapidly and technology changes are more fragmented than in the A1- and B1-scenarios. Like in the A2-scenarios population increases continuously, but at a slower rate.

The projected climate changes depend on the scenario assumed (Fig. 4.3) and vary among models. However, all models agree in projecting a temperature increase on global average. Uncertainty in projected climate changes propagates into results derived from climate model data. Lebourgeois et al. (2010), for instance, showed that the time of green-up and onset of leaf coloring depend on the climate change scenario assumed.

Although the scenarios differ related to various aspects of changes (e.g., LCC, increased GHG concentrations), they have a common reason for how these changes occur. The release of CO_2, for instance, occurs as line (e.g., traffic), area (e.g., domestic heating) or point sources (e.g., power plants, industrial complexes). Methane (CH_4), for example, is released from thawing permafrost, rise fields, and other agricultural

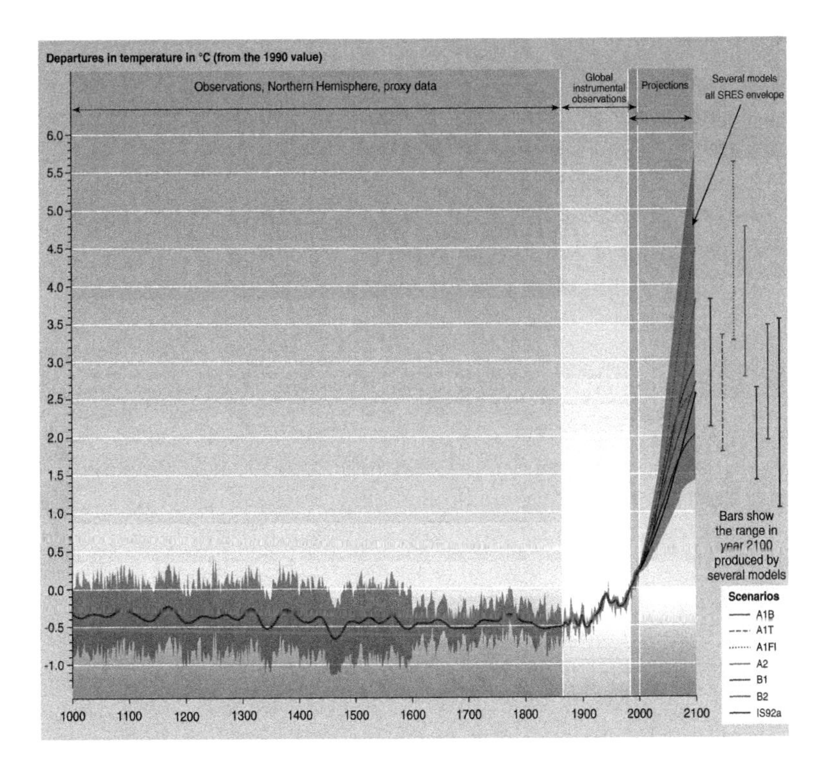

Figure 4.3. Global average temperature evolution in response to various CO_2 emission scenarios (Adapted from Climate Change 2001 Synthesis Report, p. 140)

sources. Greenhouse gases absorb and emit radiation within the infrared range with consequences for temperature wherever the GHG occur. While GHG affect the energy budget in the atmosphere directly, LCC affect only the surface-energy budget directly. In the atmosphere, LCC affect the energy budget indirectly via secondary processes.

4.2 Land-Cover Change Impacts Under Future Climate Conditions

Land-cover changes may occur in response to natural causes, a changing climate or anthropogenic actions or combinations of them. In nature, LCC and GHG concentrations change concurrently. The complex nonlinear relation between atmospheric processes inhibit to assess the concurrent impacts of LCC and increased GHG concentrations based on knowledge of the impacts of the individual changes. Thus, studies focused on LCC under future climate conditions as well as on biota changes in response to a changing climate to examine the complex biophysical feedbacks.

Some studies showed that the changes in warm-season temperatures, precipitation and evapotranspiration due to LCC can be of similar magnitude to the changes projected in response to GHG obtained for the late twenty-first century (e.g., Diffenbaugh 2009). This means that the modified biophysical land–atmosphere interactions in response to LCC can affect future climate substantially.

4.2.1 Biogeophysical Feedbacks

In Chap. 3, we discussed the biogeophysical feedbacks of LCC. Land-cover changes affect the surface-energy budget directly via altered plant physiological parameters. The energy partitioning at the surface-atmosphere interface changes in response to the modified radiative properties (e.g., albedo, emissivity). The LCC influence the latent heat fluxes via changed interception-storage capacity, leaf area index (LAI), root length, shielding of the ground, and stomatal resistance. Since the latent heat fluxes also depend on wind speed, the LCC affect the energy budget indirectly via the land-cover's aerodynamic roughness. Due to the secondary impacts of LCC and teleconnections, LCC themselves may trigger further LCC.

Climate affects the length of the vegetation season, onset of green-up, leaf coloring, and CO_2 uptake for photosynthesis. Vegetation reacts to changes in precipitation, temperature, photosynthetic available radiation, CO_2 concentration, and soil nutrients and toxics from atmospheric deposition. Once released, GHG and aerosols disperse and may have impacts far away from their sources. Under a changing climate due to GHG and aerosols release, absorption of radiation by GHG and aerosols alters the surface-energy budget. This means direct impacts on the energy budget occur wherever the additional GHG and aerosols occur. Secondary responses follow the absorption. The additional aerosols alter the availability of cloud condensation nuclei (CCN) and ice nuclei (IN). On the contrary, LCC have their remote impacts due to secondary processes and teleconnections that result as consequences of the changes in the surface-energy budget.

In the case of LCC, the modified exchanges of matter, heat, and momentum at the surface–atmosphere interface alter the temperature and moisture states of the atmosphere, and buoyancy. In the case of increased GHG and aerosol concentrations, the radiative impacts alter the temperature and buoyancy of the atmosphere. While LCC directly influence the atmospheric moisture states, GHG and aerosols influence these states indirectly. The warmer climate under increased GHG conditions enhances the atmospheric water demand. If enough soil water is available, latent heat fluxes will increase. In both cases, the altered atmospheric thermal and moisture states have consequences for cloud and precipitation formation.

A warmer atmosphere in response to increased GHG will affect the water cycle and the efficiency of the precipitation-formation process. Various studies indicated an increased residence time in response to increased GHG concentrations, that is, a "slowing-down" of the global water cycle. The increase of GHG concentrations increases the globally averaged evapotranspiration, precipitable water, and precipitation, that is, it intensifies the water cycle. Results from coupled ocean–atmosphere–land model simulations assuming the IS92A-scenario, for instance, suggested an increase of global mean near-surface air temperature, evaporation, precipitation, and runoff by 2.3 K, 5.2%, 5.2%, and 7.3%, respectively, by 2050. Like for LCC the changes in evapotranspiration, precipitation, and residence time in response to increased GHG concentrations vary in space and time.

Many Global Circulation Model (GCM) studies suggested precipitation increases by about 3.4% per degree Kelvin warming (Huntington 2006). This intensification of the water cycle can be explained approximately as an exponential increase of specific humidity with increasing temperature given by the Clausius–Clapyeron equation. The slope of Clausius–Clapyeron equation, however, suggests a 6.5% increase of precipitation per degree Kelvin. The discrepancy between the theoretical and simulated values results from the energy limitation of evaporation.

Land-cover changes may increase or decrease the atmospheric temperature and water-vapor conditions. Like the impacts of LCC on the atmosphere may diminish or enhance each other when they occur concurrently, effects from increased GHG concentrations and concurrent LCC may do so too. In midlatitudes, for instance, the conversion of natural vegetation for agricultural use may compensate for the warming due to increased GHG concentrations and increased regional precipitation (Bounoua et al. 1999).

4.2.1.1 Prescribed, Fixed LCC

Due to potential interaction of LCC and climate changes, assessment of how future LCC will affect climate is challenging. Various GCM simulations assumed different emission scenarios and LCC that persisted over the entire simulation time. The warming projected by these simulations suggests increased possibilities for agriculture in the north (e.g., Scandinavia, Alaska) and shifts of agro-ecological zones. Winter-wheat production, for instance, may occur about 90 km farther north per Kelvin warming. The yield of many crops increases under warmer conditions and increased CO_2 concentrations. History shows that increased yield leads to abandoning the least productive fields.

Few studies focused on the interaction between the atmospheric responses to increased GHG concentrations and LCC. However, biogeophysical feedbacks can enhance or diminish climate changes (Bonan 2008). Most studies suggest that LCC may have similar impacts on the atmosphere under future than current climate conditions. However, the magnitude of impact may be enhanced or diminished. Similar responses of LCC under warmer climate conditions than current climate may occur locally, while at the same time diminution or enhancement of LCC impacts may occur elsewhere, that is, interaction differs among regions. Most studies also suggest that the degree of heterogeneity of the new landscape, the size and type of LCC, and land-management remain key for the atmospheric impacts. The various state variables and fluxes show different sensitivity to the combined GHG changes and LCC.

Sánchez et al. (2007), for instance, analyzed the impact of LCC on climate under past (1960–1990) and future climate conditions (2070–2100) assuming the A2-scenario with two different land-cover scenarios. The land-cover scenarios differed only in the fraction of grassland and grass with trees in some parts of the domain. The four 30-year simulations were performed for the Mediterranean basin and most of Europe. The results for past climate, but different land-cover showed large evapotranspiration differences over areas with LCC. In some regions, the increases in evapotranspiration led to precipitation increases of up to 3 mm/d. Changes in precipitation were more complex than in evapotranspiration as the sources and sinks were more complex due to nonlocal processes and feedbacks of precipitation. The simulations with and without LCC for future climate showed similar responses to the LCC than were found for current climate conditions (Sánchez et al. 2007). Obviously, in this region the impacts of climate change are relatively independent to the land-cover descriptions.

High-resolution nested climate modeling showed that the dominant climate response to the modern nonurban land-cover distribution of the continental USA is cooling of near-surface air temperatures, especially during the warm season (Diffenbaugh 2009). Statistically significant cooling occurs regionally in the Great Plains, the Midwest, southern Texas, and the western USA. In the Great Plains, short grass changed to crop/mixed farming, in the Midwest and southern Texas noncontiguous forest changed to crop/mixed farming, and in the western USA irrigation was introduced. The related changes in both surface-moisture balance and surface albedo drive the warm-season cooling. In the Great Plains and western USA, changes in surface-moisture balance dominate, while in the Midwest changes in surface albedo dominate. In southern Texas, both effects contribute nearly similarly to the cooling. The altered surface-moisture and energy fluxes enhance moisture availability in the lower atmosphere and uplift aloft. Thus, the warm-season precipitation increases. These local and regional climate changes caused in response to the modern landscape are of a similar magnitude to those projected for future GHG (Diffenbaugh 2009).

Jonko et al. (2010), for instance, examined the impact of future global scenario–based LCC on the tropical circulation. They performed GCM simulations with present-day climate and land-cover, future climate assuming an A2-scenario with present-day land-cover, and alternatively with future land-cover. The LCC assumed base on the A2-scenario and are largest in the Tropics (36% of all land grid-cells experienced changes) and 78% of the tropical rainforests are replaced by shrub. The assumed LCC generated shallow Rossby waves trapped at the equator. These waves occurred most pronounced during boreal summer. Deforestation under A2-scenario climate conditions would shift the tropical Walker circulation patterns with an anomalous subsidence over tropical South America.

Li (2007) performed a hierarchy of 40-year simulations with the Community Climate System Model to examine LCC-climate-change interactions for pure academic CO_2 scenarios. The simulations assumed 355 ppm (reference), 720 ppm, and 1,065 ppm CO_2 conditions alternatively without and with assumed LCC in Yukon, Ob, St. Lawrence, and Colorado, and their adjacent land. These four similar-sized ($\sim 3.27 \cdot 10^6\,km^2$) regions differ by their thermal and hydrological regimes, their distance to the oceans, and their position within the large-scale circulation. Further simulations assumed the LCC just in one of the regions with and without doubled CO_2 conditions.

Doubling CO_2 increases the moisture transport from the Southern to the Northern Hemisphere during boreal summer significantly (95% confidence level) as compared to $1 \times CO_2$ conditions (Li 2007). The net moisture transport into areas north of 60 °N is delayed by a month as compared to $1 \times CO_2$ conditions. Doubling CO_2 increases precipitation and evapotranspiration for the Yukon, Ob, and St. Lawrence regions, but decreases these quantities, on average, for the Colorado region. In the warmer climate, each of the four regions exchanges more moisture with the global water cycle. The degree of interaction between the regional water cycles and the global water cycle depends on the region and season. For the Yukon, Ob, and St. Lawrence regions, small shifts in the position and/or strength of the semi-permanent highs and lows govern these changes. In the Colorado region, the increased temperature strengthens the anticyclone located over the region and the persistence of dry conditions (Li and Mölders 2008).

Under $1 \times CO_2$, $2 \times CO_2$ and $3 \times CO_2$ conditions, the regional LCC modified the regional water cycle strongly and teleconnected to areas remote from the LCC (Li 2007). The water cycle slowed down in response to doubling and tripling CO_2 regardless of the LCC. In the Yukon, Ob, and St. Lawrence regions, doubling CO_2 increased evapotranspiration and precipitation significantly in most months, while in the Colorado region, precipitation and evapotranspiration decreased. The increase in residence time due to doubling of CO_2 increases the influence of upwind conditions on the regional water cycles (Li 2007).

Due to the increased residence times under $2 \times CO_2$ conditions, LCC influence regions further downwind than under $1 \times CO_2$ conditions (Li and Mölders 2008). Occasionally, the impacts of LCC and doubling CO_2 on precipitation and residence time interact nonlinearly and significantly. Consequently, the concurrent LCC and doubling of CO_2 may

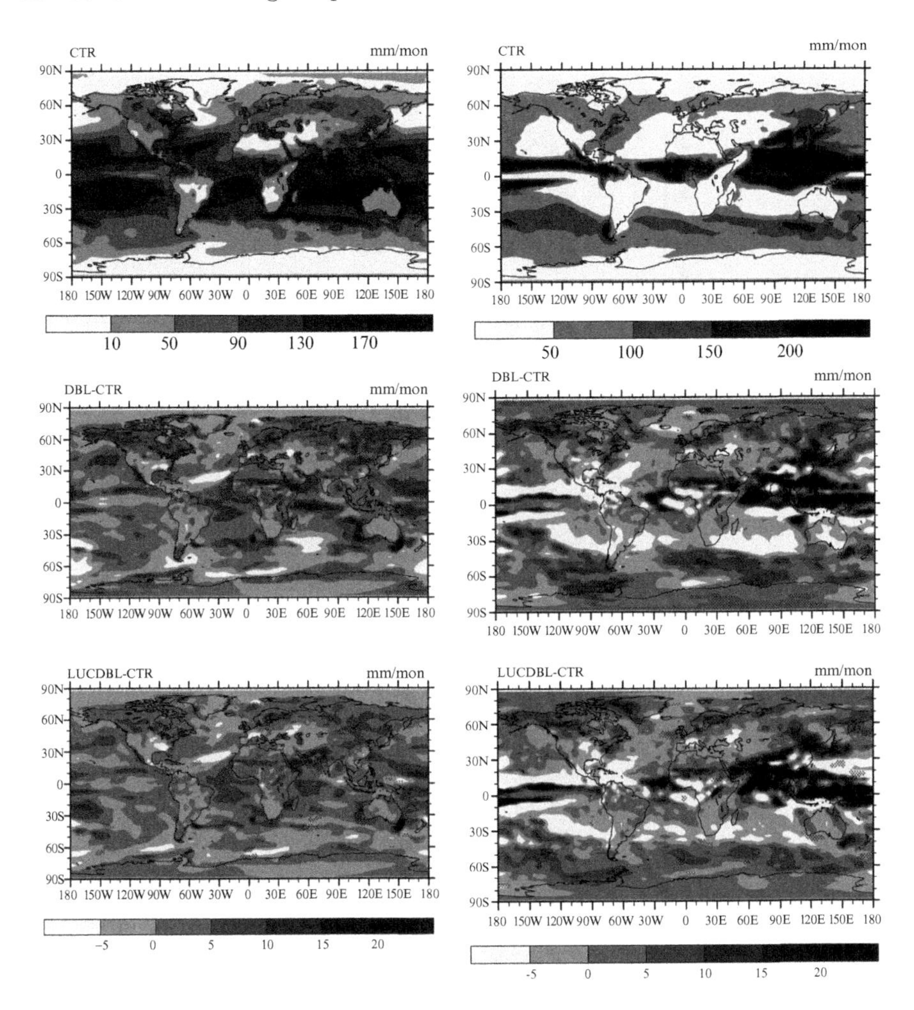

Figure 4.4. Seasonally averaged evapotranspiration (*left upper three panels*), precipitation (*right*) and runoff (*lower three panels*) as obtained for 355 ppm CO_2 concentrations (CTR) for June-July-August (JJA; *upper panel*). Differences in seasonally averaged evapotranspiration, precipitation, and runoff between conditions of doubled CO_2 (DBL) CTR (*middle panel*) and concurrently changed land-cover and doubled CO_2 (LUCDBL)-CTR (*lower panel*). Stippled areas indicate significant ($\geqslant$95% confidence level) interaction of changed land-cover and doubled CO_2 (From Li and Mölders 2008)

either enhance or diminish the individual impacts on atmospheric state variables and fluxes. These interactions vary in space and time, and are not restricted to the areas with LCC and are even visible on global average (Fig. 4.4). They result from temperature changes and the temperature sensitivity of phase-transition processes. The occurrence, duration

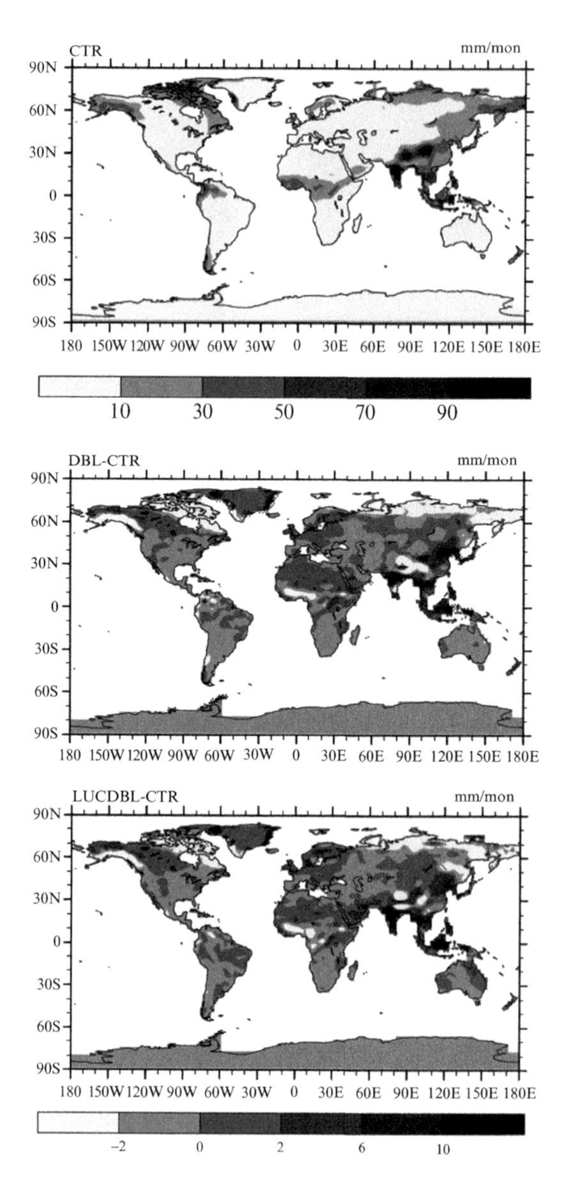

Figure 4.4. (continued)

and efficiency of (feedback) mechanism in response to anthropogenic changes depend on the season and region's climate (Li and Mölders 2008).

In all four regions, the snow-albedo feedback reduces near-surface air temperatures in winter and early spring where LCC occurred. At high latitudes, the snow-albedo feedback effects of a change from tall (e.g., forest) to low vegetation (e.g., grass) become most apparent in fall and in spring, as still enough solar radiation exists. During boreal winter, the dependency of the snow-albedo feedback on the solar zenith angle limits the snow-albedo feedback (Li and Mölders 2008).

Temporally and locally, the impacts of LCC and increased GHG may offset each other (Li and Mölders 2008). Under doubled CO_2 conditions, high-latitude winter snowfall and spring runoff increase due to the LCC. In midlatitudes, the warmer climate shifts the partitioning between liquid and solid precipitation toward the liquid phase. In the St. Lawrence region, for instance, the increased CO_2 concentrations reduced snow-cover duration, onset, and thickness. Consequently, the increased snow-albedo feedback due to LCC was effective for a shorter time (Li and Mölders 2008).

The various atmospheric impacts of LCC and increased GHG interact differently for the various state variables and fluxes. Variables or fluxes that LCC affected directly are more sensitive to interaction of LCC and CO_2 impacts than those that LCC affected via secondary processes. Evapotranspiration, for instance, is more sensitive to LCC than precipitation. Interactions between CO_2 concentration and LCC affect evapotranspiration stronger than precipitation (Li and Mölders 2008).

Under warmer climate conditions, the impacts of LCC on runoff (Fig. 4.4) differ from those under reference-climate conditions due to changes in the partitioning of precipitation between solid and liquid phase (Li and Mölders 2008). In the Yukon region, for instance, LCC hardly affect spring runoff under either CO_2 condition, while in Colorado, the LCC and warmer climate both contributed to reduced spring runoff. Doubling CO_2 increases low-level cloud-fraction and reduces long-wave radiation loss notably in regions governed by low-pressure systems and high moisture advection. The resulting strong impact on temperature may hide the temperature impact of LCC if LCC and doubling of CO_2 occur concurrently (Li and Mölders 2008).

Under $2 \times CO_2$ conditions, the LCC affect the phase and magnitude of water cycle relevant quantities in a comparable way as they did under $1 \times CO_2$ conditions in the four regions. Under $2 \times CO_2$ conditions, the assumed LCC enhanced significantly the sensible heat fluxes during summer. The enhanced temperatures and atmospheric water demands in response to the LCC increase when compared to $2 \times CO_2$ conditions without LCC. Thus, summer dryness intensifies water shortages in semi-arid regions for the assumed concurrent LCC and doubling of CO_2 (Li and Mölders 2008).

Due to the high nonlinearity of interaction between GHG and LCC impacts illustrated by the studies discussed above and various other studies, the climatic impacts of LCC should be considered when LCC are planned and/or land-use policies on grazing, irrigation, cultivation, and/or land development are made.

4.2.1.2 Gradually Changing Land-Cover

Future climate, economics, and politics drive the majority of future LCC. These LCC again affect climate and lead to further economically and politically induced LCC. In the studies discussed above, LLC were prescribed at the beginning of the simulation, that is, land-cover remained constant over the simulation period. In nature, land-cover changes gradually through vegetation dynamics except for disturbances like wildfires, landslides, volcanic eruptions, droughts, or floods and anthropogenic LCC. Biogeophysical feedback mechanisms may initiate LCC in addition to anthropogenic LCC and increases of GHG. Climate change and variability modify the length of the growing season, the temporal evolution of water availability and water amounts, and/or the amplitude of the diurnal and annual temperature cycle, and at high latitudes or high elevation, the active layer depth. These changes in local conditions put constraints on biota. Along the borders of climate zones, only slight changes in the aforementioned quantities may extinguish a biome or vegetation species from a region or permit a biome or vegetation species to invade a region in which it previously was unable to survive. Changes in what to cultivate and where may be required to cope with the new climatic conditions.

Ecological responses to climate change modify the biogeophysical feedbacks of forests to climate that vary with latitude. In boreal forests, for instance, temperature increases enhance the net primary production (negative feedback). Expansion of boreal forests reduces the surface albedo and may trigger further warming. Thus, the boreal forest expands further north into the tundra and dies at its southern flank (Bonan 2008).

In tropical forests, on the other hand, increased temperatures enhance the evaporative demand (Bonan 2008). Consequently, the soil dries. The net primary production goes down (positive feedback) because of changes in stomatal conductance, LAI, and, on a long term, species composition. The increased atmospheric CO_2 concentrations reduce stomatal conductance and evapotranspiration thereby triggering further warming. Simulations with a climate model interactively coupled with an ecosystem model suggested that this positive climate feedback may lead to loss of tropical forests (Bonan 2008).

To examine the interaction between climate and vegetation dynamics Watson and Lovelock (1983) developed a simple model known as "Daisyworld." On a Daisyworld cloudless planet, only black and white daisies exist. They respond to climate via temperature. The black daisies reflect less solar radiation than the bare ground, while the opposite applies for the white daisies. Thus, the black daisies have a warming and the white daisies have a cooling effect on climate. White (black) daisies are favored by warmer (cooler) mean temperatures. The growth rate depends parabolically on temperature. These conditions have a stable point around which the daisy community homeostats temperature over a wide range of luminosities.

In a Daisyworld with clouds, clouds form over black daisies as air rises over warm areas. Under these climatic conditions, more black daisies mean more clouds and a cooler planet as clouds reflect radiation. The white daisies that are less well adapted to cooler temperatures become extinct. The black daisies achieve finally an effective homeostasis (Watson and Lovelock 1983).

The results of the simple Daisyworld illustrate that nonlinear feedback mechanisms are at work and less-adapted species will become extinct. The very temperature-sensitive growth rate feeds back on any change and buffers the stable state against external variations (Watson and Lovelock 1983). All biota has a peaked growth versus temperature curve and thus may influence the Earth's temperature substantially. The complexity of the real world makes it often difficult to assess what feedbacks may be triggered and how feedbacks interact (enhance or diminish) with each other. To examine the feedbacks between biogeophysical effects and anthropogenic climate change, vegetation models are required that are run inline with the climate model.

Vegetation models use many empirical parameters to represent the vegetation changes in response to climate changes. These parameters bear high uncertainty that propagates into the projection. Thus, various studies focused on the sensitivity of projections to these parameters. Hallgren and Pitman (2000), for instance, found that parameters affecting photosynthesis, evapotranspiration, and root distribution influence the simulated plant functional types (PFT) distribution the most. These parameters affect the net primary productivity of the PFT. The net primary productivity determines the balance among PFT or between C3 and C4 plant types, and their global distribution shifts. The competitive balance between grass and trees is very sensitive to the evapotranspiration and root-distribution parameters. Since the uncertainty of parameters propagates into the state variables and fluxes, scientists must choose these parameters very carefully. Further measurements are required to determine these parameters for the various vegetation types.

Inclusion of vegetation dynamics models may account for feedback processes that otherwise would be ignored (Bonan 2008). As discussed in Chap. 3, the higher albedo of crops permits for a cooler climate than temperate forests. The often high evaporative cooling from irrigated crops may further contribute to cooling. During droughts, forests maintain their greenness because their long roots have access to deeper soil layers or even groundwater, while crops lose their greenness. Consequently, the sensible heat fluxes and near-surface air temperatures increase up to 13 K stronger over crops than forests. Furthermore, the impacts of deforestation on climate differ for deciduous, mixed, and coniferous forests (Bonan 2008).

Studies on climate-induced changes of biomes and their influence on climate indicate coherent regions of substantial changes in precipitation, near-surface air temperature, humidity, and wind speed as the outcome of meso-α-scale LCC. The big changes in these state variables and fluxes are related closely to the changes in the vegetation parameters. Neglection of vegetation feedback mechanisms may provide locally changes of opposite sign. Alo and Wang (2010), for instance, demonstrated the role of vegetation dynamics for future climate conditions of western Africa. They used a dynamic vegetation model coupled asynchronously to a regional climate model. They performed simulations for the twentieth century and future climate (A1B) with the coupled regional climate–vegetation model and with the regional climate model alone wherein vegetation distribution remained at present-day state. This suit of simulations permits investigating the impact of structural vegetation feedback on hydrological processes and simulated climate. They found that over large parts of West Africa, vegetation feedback on future climate changes exceeds or is of similar a magnitude to CO_2 induced radiative and physiological effects during June-July-August (JJA). The simulations with the regional climate–vegetation model suggest widespread increase in LAI and significant shifts in vegetation type. Compared to 1984–1993, the Sahel experiences a 31% increase of grass cover, while the Guinean Coast experiences a 56% increase of drought deciduous tree coverage and a 49% reduction of evergreen tree coverage in 2084–2093. These vegetation changes feed back to a 23% increase of rainfall over the Sahel, and a 2% decrease over the Guinean Coast (Fig. 4.5). Ignoring the vegetation feedback mechanism, that is, just considering the radiative and physiological effects of higher atmospheric CO_2 concentration led to a 5% decrease of JJA rainfall in both regions (Alo and Wang 2010).

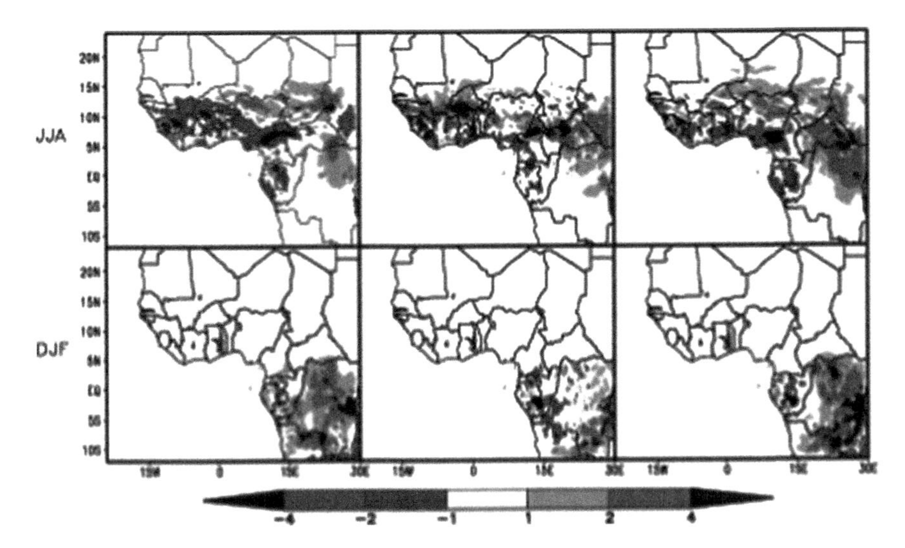

Figure 4.5. From *left* to *right* changes in rainfall (mm/d) due to the radiative and physiological effects of CO_2 alone, vegetation structural changes alone, and the combined impact of vegetation structural changes and radiative and physiological effects of CO_2 during JJA (*upper panels*) and DJF (*lower panels*) (Modified after Alo and Wang 2010)

Radiative and physiological effects in response to increased CO_2 lead to warming everywhere during the wet (JJA) and dry (December-January-February, DJF) seasons. Near-surface air temperatures increase by 2.5 (2.9) and 3.4 (2.8) K over the Guinean Coast and Sahel, respectively, in JJA (DJF). Structural vegetation feedbacks diminish the JJA warming by 6% and 27% along the Guinean coast and in the Sahel, respectively. They enhance it by 3% and 11% in these regions, respectively, in DJF due to reduced albedo. During the dry season, the albedo effect dominates because drought deciduous trees are in senescence, that is, evaporative cooling is negligibly small. The combined CO_2 and radiative and physiological effects reduce JJA evapotranspiration and increase soil moisture in the Sahel. Including vegetation feedback effects increases both quantities in the Sahel in JJA (Alo and Wang 2010).

Under future climate conditions, the atmospheric responses to LCC remain sensitive to large-scale circulation patterns like the North Atlantic Oscillation (NAO), Pacific Decadal Oscillation (PDO), and El-Niño Southern Oscillation (ENSO) (e.g., Li and Mölders 2008; Ramos Da Silva et al. 2008). Ramos Da Silva et al. (2008), for instance, performed simulations with the Regional Atmospheric Modeling System (RAMS) to examine the impact of deforestation in the Amazon basin for

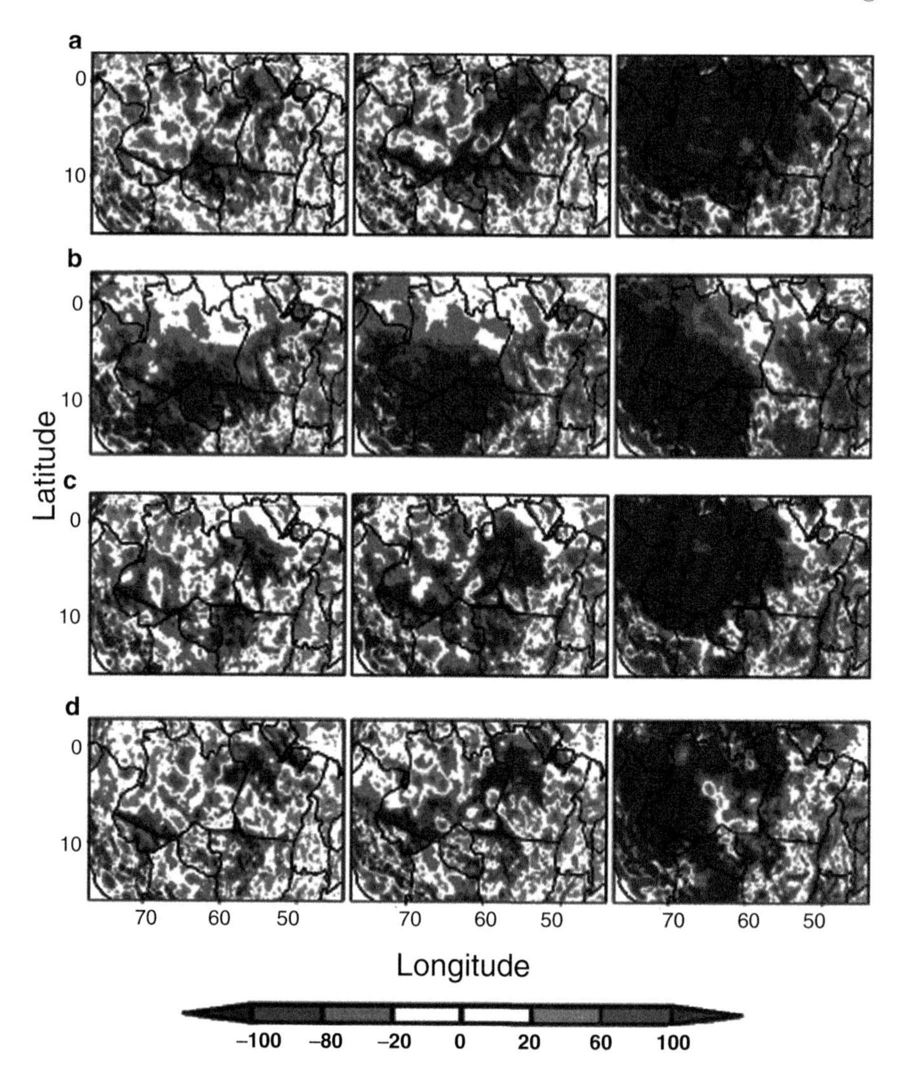

Figure 4.6. Rainfall anomaly (mm) relative to the reference simulation for the LCC scenario of 2030 (*left*), 2050 (*middle*), and total deforestation (*right*) for the meteorological conditions of (**a**) 1997, (**b**) 1998, (**c**) 1999, and (**d**) 2000 (Modified after Ramos Da Silva 2008)

2030 and 2050 under various climate conditions. Like for past and current climate, the basin-averaged rainfall decreases progressively, as the deforested area grows with time. Precipitation increases notably at the edge of the massively deforested area and at elevation. Over the deforested area, precipitation decreases notably and squall lines propagating through dissipate before reaching the western part of the basin (Fig. 4.6).

As deforestation progresses, precipitation changes more distinctly during El Niño events than during wet years (Ramos Da Silva et al. 2008).

Land-cover not only changes in response to climate variability and change but also due to fire disturbance. In general, fire risk varies dramatically over a short time in response to previous weather, frequency of lightning, and human activities. Lightning strikes are mainly responsible for the occurrence of wildfire clusters (i.e., several large fires occurring simultaneously in the same region). Periodic extreme fire-weather situations ensure that wildfire clusters initiate extended LCC in fulfillment of their essential ecological role.

Various GCM studies suggest that fire danger increases in a warming climate (Soja et al. 2007). Climate and biogeophysical changes that enhance the frequency of thunderstorms and/or the strength of their updrafts affect the number of lightning strikes and number of wildfires. Lightning activity depends on charge transfer and separation, which mainly occurs during graupel formation in the updrafts of thunderstorms. Since the LCC related to wildfires enhance the likelihood of severe convection, these biogeophysical feedbacks may lead to further wildfires and wildfire-induced LCC as well (Mölders and Kramm 2007).

As discussed in Chap. 3, in high latitudes, fire-induced LCC affect surface albedo also in winter when the fire scar is snow covered. Moreover, in high latitudes, climate change and variability not only affect fire regimes, but also snow-cover duration. The modifications due to fire-induced LCC yield to local heating/cooling of the atmosphere via the temperature-albedo feedback. Euskirchen et al. (2009), for instance, examined for the A2-, B2-, and two other climate change scenarios the vegetation changes due to changing fire regimes and snow-cover duration in the western Arctic for 2003–2100. Forest-stand age affects summer temperatures. Thus, the future fire regimes modify the forest-stand age distribution. The related increases in summer albedo lead to marginal cooling. At the same time, the related decreases in snow-cover and length of the snow season (4.5 days/decade) reduce winter albedo. Consequently, winters are slightly warmer. On annual average, the changes result in a net warming (Euskirchen et al. 2009). However, the associated fluxes (net over all simulations amounts 3.4 W m^{-2}/decade) are below current accuracy of surface-flux measurements (±5% for net radiation of the measured reading when ventilated, ±10% of the measured reading for soil-heat flux, sensible and latent heat fluxes). Thus, measurement technique has to be improved to evaluate the findings.

4.2.2 Biogeochemical Feedbacks

Most climate models consider the atmosphere and biosphere as a coupled system with respect to biogeophysical processes only. However, in addition to affecting the core biogeophysical processes (e.g., surface-energy fluxes, hydrologic cycle) LCC influence the trace gas cycle with consequences for vegetation dynamics. This means that biology may adapt to climate and atmospheric composition and thereby feed back to climate and the atmospheric composition. Vegetation affects the atmospheric composition, for instance, by consumption of CO_2 during photosynthesis. Increasing CO_2 concentrations act as fertilization enhancing the photosynthetic activity, which feeds back negatively to higher CO_2 concentrations (Bonan 2008).

The biosphere influences and to a certain degree regulates atmospheric composition, chemistry, and climate via biogeochemical feedback processes (Arneth et al. 2010). These interactions between the geochemistry and biota of a region link the trace gas, water, and energy cycles. Observations evidence biogeochemical responses also to anthropogenic air pollution and human-induced LCC. Afforestation, for instance, sequesters carbon and attenuates warming, meaning a negative climate forcing (Bonan 2008).

Biogeochemical processes modify the distribution of the relatively well-mixed GHG that affect the radiation budget. The changed atmospheric composition influences the biogeochemical processes of radiatively active gases. The related climate change modifies the biogeochemistry of non-radiatively active compounds that affect the concentrations of GHG. The altered atmospheric composition modifies the wet and dry deposition of atmospheric contaminants into ecosystems. Depending on the species, wet and dry composition can be either a burden or a fertilizer for the ecosystems. The modified deposition may further affect climate as the potential CO_2 impacts on plant biomass depend on the availability of nutrients and water (Arneth et al. 2010).

The carbon cycle may strongly affect the impact of LCC on the atmosphere. Hints exists that overall, the impact of biogeochemical feedbacks on climate could be similar in magnitude to the physical feedbacks in the climate system; the magnitude of individual processes and the synergies between them are still under research (Arneth et al. 2010). Most studies on biogeochemical feedbacks to warming over the twenty-first century suggest that, on average, ecosystem changes resist warming mainly through ecosystem carbon storage, that is, net negative feedbacks (Field et al. 2007).

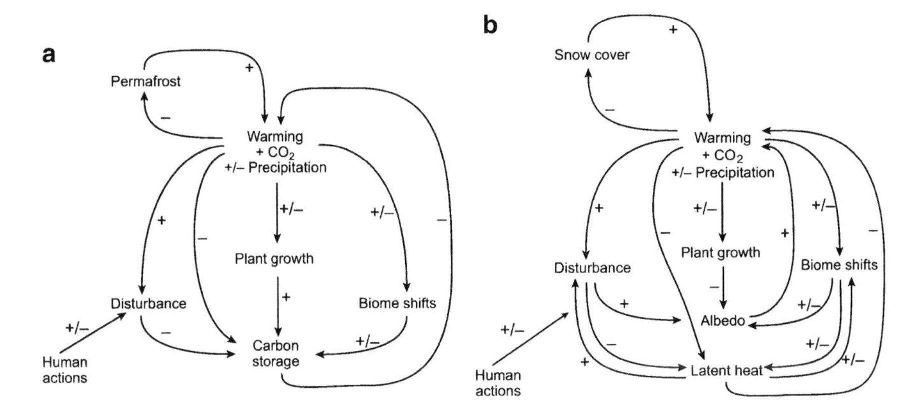

Figure 4.7. Schematic view of major processes involved in the control of (**a**) CO_2 feedbacks and (**b**) biogeophysical feedbacks in the climate system. The symbol $+(-)$ indicates that increases in one process lead to increases (decreases) in the next. The set of forcing processes (plant growth, biome shifts, disturbance) is mostly similar the biogeochemical feedback via CO_2 and the biogeophysical feedbacks (From Field et al. 2007)

Figure 4.7 schematically illustrates carbon feedbacks. The degree of climate warming determines how effective ecosystems feed back to climate. Under modest emission scenarios, for instance, negative and positive net feedbacks of terrestrial ecosystems occur at low and high latitudes, respectively. High-emission scenarios lead to positive feedbacks. At high latitudes, negative feedbacks due to forest expansion may be compensated by positive feedbacks from reduced albedo, and increased carbon emissions from wildfires and permafrost thawing (Field et al. 2007).

These studies suggest that realistic description of biogeochemical and biogeophysical processes is important for future projections and planning of resources. However, examining the complex biogeochemical feedbacks requires earth-system models that represent biogeochemical processes, anthropogenic LCC, and vegetation dynamics in addition to the biophysical processes (Fig. 4.8). Such models must permit to describe the response of plant ecosystems to GHG- and/or LCC-induced climate change. The scientific community started efforts on coupling climate models with dynamic vegetation and chemistry models toward the goal of assessing the importance and role of biogeochemical feedbacks for climate. The results of first efforts demonstrated that LCC and related biogeochemical and biogeophysical processes may substantially affect future climate. There are indications that the interaction

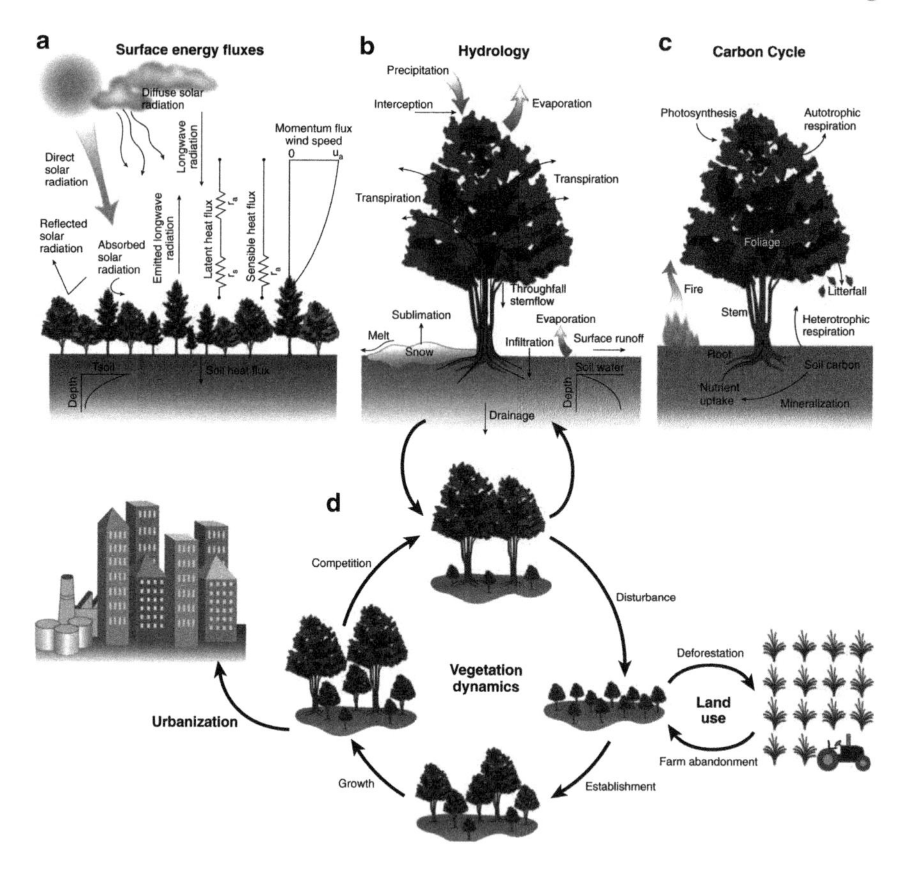

Figure 4.8. Schematic view of (**a**) biogeophysical processes, (**b**) the hydrological cycle, (**c**) carbon cycle, and (**d**) anthropogenic and vegetation dynamical response to LLC and climate change–induced changes (Modified after Bonan 2008)

of biogeochemical and biogeophysical processes may locally even offset GHG emission impacts. However, due to the huge computational demands of such models the number of studies is very limited. The models vary in their complexity and their results differ widely depending on the degree of complexity considered. However, all these studies document the potential importance of biogeochemical processes for climate.

Claussen et al. (2001), for instance, examined the concurrent impacts of climate-induced LCC considering biogeophysical and biogeochemical feedbacks on climate with a coupled atmosphere-biosphere-ocean model of intermediate complexity. The increase in atmospheric CO_2 and radiative forcing due to tropical deforestation outweighs the biogeophysical effects and warms the climate. In mid and high latitudes, the snow-vegetation-albedo feedback in synergism with the sea-ice-albedo feedback outweighs the biogeochemical processes. Thus, deforestation in these regions cools and afforestation warms the climate on global average. On the regional scale, a complex pattern of warming and cooling is established.

Bonan (2008), for instance, examined the impacts of likely LCC under the socioeconomic environmental conditions of the A2- and B1-emission scenarios. The projections with an A2-scenario assumed widespread agricultural expansion with farming on the most suitable land for agriculture by 2100. The projections with the B1-scenario assumed land abandonment and afforestation in consistence with the storyline of a decline in population and improved agricultural efficiency. Bonan (2008) performed these simulations without and with consideration of the carbon cycle. Under the A2-scenario and its assumed LCC and exclusion of the carbon cycle, the forest loss enhances the warming in Amazonia. The cooling due to the more intensive agriculture reduces the warming in midlatitudes. Under this B1-scenario and its assumed LCC and exclusion of the carbon cycle, temperatures increase by 1 K where no or only small LCC occur. Inclusion of the carbon cycle led to similar global average climates for the twenty-first century for both scenarios despite their different socioeconomic storylines. In the A2-scenario with LCC, the increased agricultural use of land yields biogeophysical cooling. On the contrary, in the B1-scenario with LCC, the reforestation of temperate forest leads to biogeophysical warming. The net carbon loss due to deforestation leads to greater biogeochemical warming for the A2-scenario with LCC than the B1-scenario with LCC. This warming results from the extensive deforestation in the A2-scenario and the temperate afforestation and lower tropical deforestation in the B1-scenario. In the A2-scenario with LCC, the biogeochemical warming diminishes the biogeophysical cooling leading to a net warming on global average. In the B1-scenario with LCC, the moderate biogeophysical warming from temperate afforestation adds to the weak biogeochemical warming from tropical deforestation. The combined feedbacks result in a net warming on global average similar to that obtained in the A2-scenario with LCC (Bonan 2008)

Schurgers et al. (2008) performed a suit of simulations with an earth-system model to examine feedbacks between the climate and biosphere. Their ensemble simulations assumed CO_2 emissions according to historical data and the B1-, A1B-, and A2-scenarios with an exponential decay after 2100. The feedback mechanisms lead to maximum atmospheric CO_2 concentrations of 520 ppm (B1), 860 ppm (A1B), and 1,680 ppm (A2) between 2200 and 2500. To separate the impacts of biogeophysical, biogeochemical, climate, CO_2, and LCC, they performed the simulation with and without inline vegetation model. Furthermore, they performed a simulation with CO_2 emissions and suppressed climate change, and a simulation with prescribed land-cover. The reduced albedo at high latitudes was the biogeophysical change with the largest influence on climate. By 3000, the reduced albedo augmented the temperature increase of 1–2 K in some high-latitude areas in the A2-scenario. However, these changes are small compared to the warming caused by the increased CO_2 concentrations (Schurgers et al. 2008).

Current modeling of carbon storage shows high uncertainty that makes it difficult to assess the impact of future LCC on carbon storage. Biospheric uptake of CO_2 varies temporally and spatially and among scenarios between 15% and 30% of the CO_2 emissions. The Tropics and Subtropics store carbon fast. At high latitudes, however, the expansion of boreal forest, among other things, slows down the carbon storage (Schurgers et al. 2008).

Most studies restricted biogeochemical feed backs to carbon only. However, interactions between the nitrogen and carbon cycles may stimulate or limit carbon sequestration (Arneth et al. 2010). Such biogeochemical feedbacks may diminish or even eliminate the cooling effect due to CO_2 fertilization. Depending on the strength of interaction, the total positive radiative forcing due to biogeochemical and biogeophysical feedbacks may reach between 0.9 and 1.5 $W\ m^{-2}K^{-1}$ by the end of the twenty-first century (Arneth et al. 2010).

Another uncertainty of future LCC impact assessment results from changing fire regimes and related feedbacks. Boreal ecosystems store a large amount of carbon in the soil, permafrost, and wetlands (Bonan 2008). A shift toward younger stands (i.e., higher fraction of deciduous forest) due to more wildfires or deforestation increases albedo, especially in winter. The temperature-albedo feedback may offset the forcing from GHG and result in cooling. The climate forcing from younger stands may be of similar magnitude than that of ecosystem changes. The gradual post-burn changes of the landscape alter the balance between the increase in surface albedo and the radiative forcing from GHG changes. On average over an 80-year fire cycle, the negative forcing due to the surface albedo increases exceeds the positive biogeochemical forcing (Bonan 2008).

The discussed and various other studies suggest that a future challenge is to develop earth-system models capable to describe the biogeochemical and biogeophysical processes in reasonable computational time. Another challenge is to design plausible combined anthropogenic LCC and emission scenarios that reflect future conditions as realistic as possible to reduce uncertainty. Any assessment or adaptation modeling using climate-projection data as input will propagate the uncertainty of the data used for the projections. Methods have to be developed on how to deal with uncertainty in climate assessment and adaptation planning.

4.3 Air Quality

Assessment of the concurrent impacts of GHG and LCC-induced climate change on air quality is rather difficult. Thus, assessment of future air quality faces challenges on various fronts.

Today's thresholds of the National Ambient Air Quality Standards (NAAQS) are based on current best medical knowledge. It is hard to foresee changes of the NAAQS and which species may be found to be hazardous/health adverse as medical sciences advance.

The computational demands of climate and photochemical simulations including the interaction of the trace, water, and energy cycles are very high. Land-cover and climate changes affect chemical transformation via modified temperature, relative humidity, insolation, removal rates, and oxidative capacity. Changes in cloud characteristics (e.g., liquid and ice water content, height of cloud base and top, albedo, lifetime, time of cloud formation and dissipation); cloud frequency, and cloud and precipitation distributions affect the equilibrium between the aqueous and gas phase; gas, aqueous, and aerosol-phase reactions; and the removal of pollutants from the atmosphere as well as photolysis rates. An increase in temperature and humidity, for instance, may reduce the net ozone concentration, on global average, but increase the ozone concentrations in cities. Changes in temperature, humidity, and insolation affect the biogenic emissions (e.g., monoterpene, isoprene). Despite atmospheric chemical reactions are not energetically relevant, the altered distribution of atmospheric species may affect the energy budget directly and indirectly. Such changes in the energy budget modify the fields of temperature, wind, and precipitation.

Another serious difficulty is the uncertainty in the LCC, emission rates, and emitted species and their spatial and temporal distribution. Changes in energy sources, advances in technology, migration into cities, and shifts in transportation habits affect the emission profiles. Urbanization and industrialization increase the number of point sources and shift emissions from biogenic to anthropogenic. In a warmer climate, in addition to altered biogenic emissions from LCC, emissions from domestic heating go down, while those from air conditioning go up.

Due to all these complications, studies on future air quality typically concentrate on an aspect of LCC on future air quality. Most of them focus on a specific region and limited period (e.g., season) with specific questions in mind. Chen et al. (2009), for instance, assessed the impacts of climate change on future biogenic emissions, O_3 concentrations and biogenic secondary organic aerosols (SOA) in the USA. They performed simulations with a regional modeling system for five summer months for 1990–1999 (case 1) and three future cases. Their case 2 scenario assumed present-day land-cover for 2045–2054. Case 3 used projected future land-cover for 2045–2054, and case 4 assumed future land-cover with designated forest regions to sequestrate carbon.

The land-cover projection plays an important role in assessing future regional air quality. Projected biogenic emissions depend on the projected regional climate and LCC (Chen et al. 2009). The increased temperature and solar insolation enhance the average isoprene and monoterpene emission rates by 26% and 20% even without LCC (case 2). Concurrent LCC and climate changes (case 3) decrease isoprene emissions by 52% (Fig. 4.9) and monoterpene emissions by 31%. The decrease results primarily from the projected expansion of cropland. Regional afforestation together with climate changes reduce isoprene and monoterpene emissions by 31% and 14%, respectively, as compared to case 1. Nevertheless, the average daily maximum 8-h O_3 concentrations increase between 8 and 10 ppb in case 3 and 4 due to the increase in future anthropogenic emissions and the change in oxidation capacity (Fig. 4.9). Compared with case 1, the future LCC scenarios led to spatially varying ozone differences of ±5 ppb (Chen et al. 2009).

The biogenic SOA concentrations responded directly to the changes in monoterpene emissions. In response to the warmer climate, SOA concentrations increased by 8% for present-day land-cover, while they decreased between 28% and 45% for the future LCC (Fig. 4.9). Regionally, LCC can offset temperature-driven increases in biogenic emissions (Chen et al. 2009).

The fraction of SOA of the total aerosols may increase under warmer climate conditions, but certain LCC may reduce the SOA burden. Heald et al. (2008), for instance, examined the sensitivity of SOA concentrations to changes in climate and emissions for the A1B and A2 scenario. Under these storylines, global sulfur emissions decrease. The authors included yields for SOA formation from monoterpene oxidation, isoprene photo-oxidation, and aromatic photo-oxidation in the Community Atmospheric Model version 3 (CAM3). The Model of Emissions of Gases and Aerosols version 2 (MEGAN2) within the Community Land Model version 3 (CLM3) coupled to CAM3 calculated the biogenic emissions of isoprene and monoterpenes.

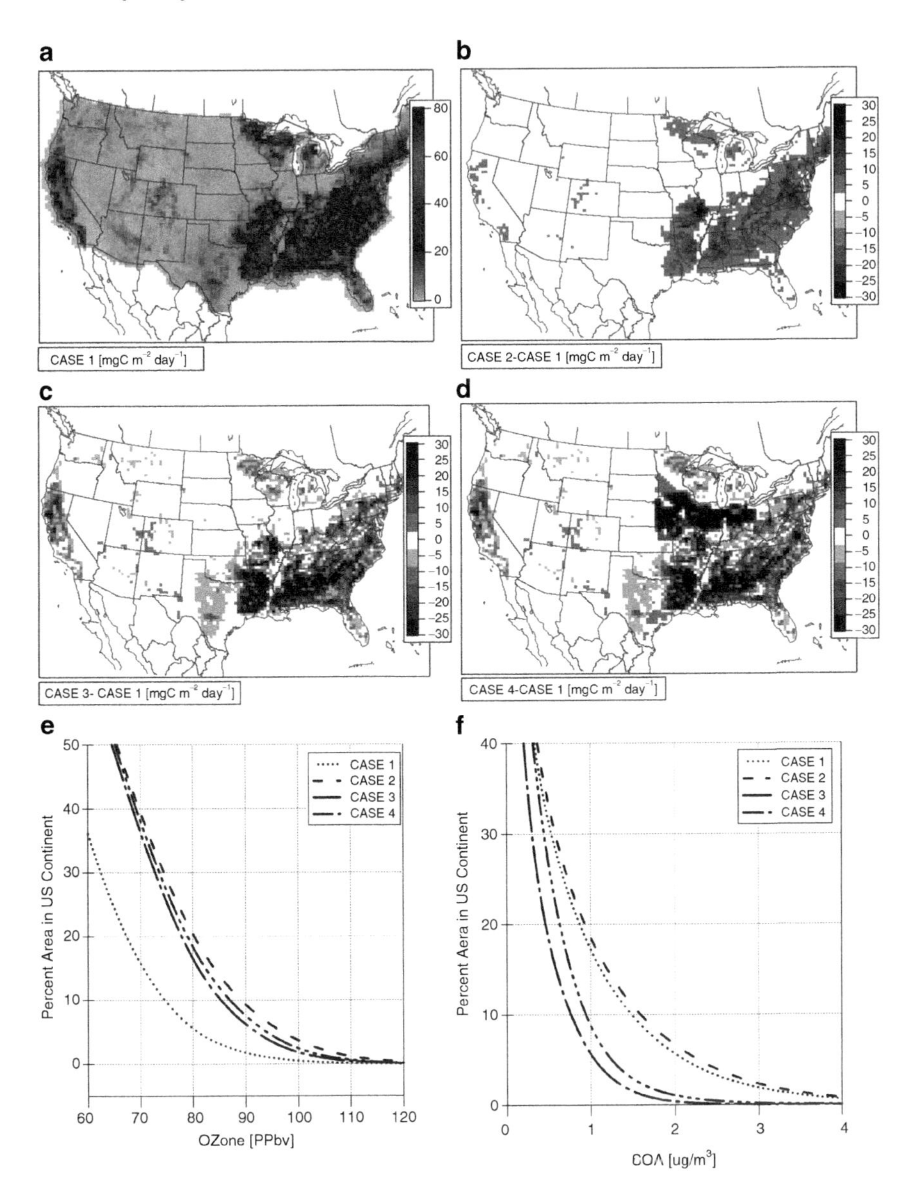

Figure 4.9. Mean biogenic isoprene emission rate for the present-day for (**a**) case 1 and differences in isoprene emissions for the future cases minus present-day case for case (**b**) case 2, (**c**) case 3, and (**d**) case 4. Percent area cumulative distribution of (**e**) daily maximum 8-h ozone mixing ratio and (**f**) 24-h biogenic secondary organic aerosol (SOA) concentration (Modified after Chen et al. 2009)

Heald et al.'s (2008) simulations suggested that the largest current and future source region for SOA is South America. By 2100, the largest relative growth in SOA production occurs in Asia as large increases in anthropogenic aromatic emissions are projected here. On global average, the increase of biogenic and anthropogenic emissions enhances the SOA concentration by 26% and 7% in 2100 in the A1B scenario. In total, the SOA concentration increases by 36%. Note that the increase due to anthropogenic SOA also includes enhanced biogenic SOA formation related to increased primary organic aerosol emissions. Globally, the SOA-burden increased by 6%. The results with projected anthropogenic LCC for 2100 (A2) showed a reduction of the global SOA burden by 14%. Heald et al. (2008) attributed this reduction mainly to the expansion of cropland.

4.4 Food and Fuel Production

According to projections, the global population will exceed eight billion by 2025. Providing adequate amounts of food at affordable prices for everyone is one of the future challenges of the world's agriculture. If demands for biofuel increase, more agricultural areas will serve for biofuel production.

Most fertile land suitable for agricultural production already serves this purpose. In many regions, the potential to expand agricultural production is irrigation limited. Typically, new lands converted for agricultural use are fragile, and often have short productive lifetime. Studies for Amazonia, for instance, show that converting land for agricultural use can have serious negative long-term effects on weather and climate, and may affect yields in the region. Agricultural production on land of low suitability for this purpose requires intensive use of pesticides and fertilizers. These chemicals may pollute ground and surface water and the atmosphere and may have adverse effects on food quality, health, and safety.

Enhanced agricultural production affects climate also via altered GHG and biogenic emissions. An increase of rice production, for instance, increases CH_4 emissions. Efforts to reduce CH_4 emissions from rice fields by altering the water availability affect the surface-energy balance. These biophysical feedbacks and related secondary changes affect climate in a different way than CH_4 did. Thus, a future challenge is to examine trade-offs and benefits of efforts associated with LCC and reducing GHG emissions.

Another challenge is the behavior of plants under and adaptation to altered CO_2 conditions. Studies conducted under controlled environmental conditions (e.g., chamber measurements) and/or optimal field conditions suggest that most plants increase their photosynthesis rates under increased atmospheric CO_2 conditions (Kimball et al. 2002). In addition, the enhanced CO_2 concentrations reduce the stomatal openings for some crops. This behavior reduces the transpiration per unit leaf area and improves the water-use efficiency (defined as the ratio of crop biomass to the amount of evapotranspirated water). Consequently, the growth and yield of most agricultural plants increase as CO_2 concentrations increase (Kimball et al. 2002). Further research is needed to understand this behavior, and response functions including controlling parameters have to be determined.

A main future challenge is to provide enough food for the growing population under changing climate and land-cover conditions. The SRES scenarios assume nonlinearity for the relationship of food production and altered climate and land-cover. Studies with these scenarios suggest that challenges depend on the future emission behavior and will differ with regions with potential for conflicts. Parry et al. (2004), for instance, examined directly the potential influence of climate change on food production with the HadCM3 global climate model. They used transfer functions derived from crop-model simulations to assess the future yield assuming the A1FI-, A2-, B1-, and B2-scenarios. Their results showed that in developed countries, crop production mostly benefits from climate change. The global crop yield seems to remain stable, but with regionally notable changes. The regional differences increase with time, especially under scenarios of great inequality (A1FI, A2). The A1FI scenario provided the greatest regional and global decreases in yields (Fig. 4.10). The A2a-c scenarios provided the largest differences in crop yield between developed and developing countries. Developed and developing countries experience similar changes in crop yield under the B1- and B2-scenarios, but with slightly greater changes in crop yield for the B1- than B2-scenario (Parry et al. 2004).

4.4.1 Water Availability

Water availability at the right amount at the right time and place is critically important for food and biofuel production as well as human activities and welfare. Climate- and LCC-induced changes in solid precipitation pattern may modify strongly the annual runoff pattern (e.g.,

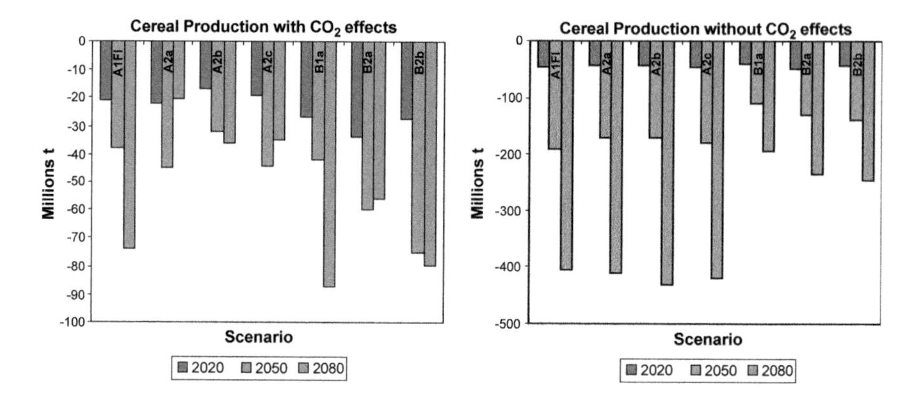

Figure 4.10. Changes in global cereal production due to anthropogenic climate change for seven SRES scenarios with and without consideration of CO_2 effects relative to the reference simulations that assumed no climate change (Modified after Parry et al. 2004)

Fig. 4.4), even if precipitation intensity and amount stay the same. Thus, GHG- and LCC-induced climate fluctuations can increase the pressure on potable, agricultural and industrial water supply particularly in drylands, areas frequently facing droughts, and areas depending on glaciers and/or seasonal snow-packs for water supply during the warm season.

More than one sixth of the world's population relies on water from seasonal snow-packs and/or glaciers. The warmer climate in response to various GHG scenarios (Barnett et al. 2005) would affect the hydrological cycle strongly, especially in these regions. In a warmer climate than today, the ratio of solid and liquid winter precipitation shifts toward the liquid phase. Less snow accumulates over winter, while winter runoff increases. Furthermore, snowmelt sets on earlier in spring than under current climate. Changes in the temporal availability of precipitation may require irrigation and/or changes in agricultural production to adjust to and/or make the best out of the water available at a given time.

Water is available as green or blue water. Blue water refers to surface water from rivers, lakes, or water reservoirs, and groundwater and is directly available for urban, industrial, or agricultural use. Huge fractions of irrigation use virgin blue water. This source has only limited capacity and once an aquifer is emptied other water sources have to be found. Green water refers to the fraction of precipitation stored in the soil that plants can take up for growth (plant-available water). Plant-available water is given by the difference between the soil-water content at field capacity and at the permanent wilting point. The permanent wilting point depends on the soil type.

Soils having large values of potential plant-available water are more suitable for agricultural purposes than those with low potential water availability. Agricultural practices (soil densification due to use of heavy equipment, wind erosion, etc.) may affect the maximum plant-available water. Any soil degradation may affect evapotranspiration strongly during droughts.

Under warmer climate conditions, vegetation has increased water demands. Evapotranspiration may increase at least locally with potential needs for irrigation to sustain productivity. Even if precipitation remained the same, the altered partitioning of precipitation between infiltration, evapotranspiration, and runoff reduce runoff. Changes in the temporal evolution of runoff may require constructing or modifying storage capacities (e.g., dams, artificial lakes) to guarantee sufficient water availability year round for the growing population, hydropower production, and/or to avoid seasonal flooding. Such measures mean LCC with consequences on at least local climate.

The results of many GHG scenario projections indicate substantial changes in the temporal and spatial pattern of precipitation amount and intensity. Such changes alter the temporal evolution of runoff and runoff volume. If rivers discharge through several countries or water reservoirs are divided between several states, a big future challenge is to coordinate water management especially during droughts. Water-management agreements have to be negotiated. These agreements will have consequences for land-use and at least locally for atmospheric conditions.

In high latitudes, changes in precipitation in response to LCC and/or climate change provide challenges for residents, rivers, ecosystems, and the economy. Increased summer precipitation increases the risk for flooding in mainly rain-fed drainage basins. Changes in solid precipitation amount affect frost-depth with implications for the onset of green-up, river-water levels during breakup, survival of ecosystems, and potential for agriculture. Thinner snow-packs mean deeper frost-depth that may diminish the increase in length of the growing season due to the warmer climate. Changes in winter snowfall and snow-depth affect the onset and intensity of the fire season and/or outdoor winter activities and tourism. All these changes influence the economy.

Under present climate, several of the world's large lakes and wetlands have a delicate balance between inflow and outflow. Such watersheds are highly vulnerable to climate changes and LCC. An increase of evapotranspiration by 40%, for example, could result in much reduced outflow, and subsequent drying and desertification (Sivakumar 2007). These watersheds require appropriate adaptation strategies to prevent floods and sustain their ability to supply water, food, and services adequately.

4.4.2 Soil Degradation and Land Loss

Various climate-change scenarios suggest that climate variability, frequency, and lengths of droughts will increase in the Mediterranean Basin, much of interior Australia, and the Southwest of the USA (Verstraete et al. 2009). The Hadley cells may intensify and/or extend poleward. The increased aridity in these subtropical regions may enhance desertification and soil degradation with the consequences for land-cover and at least local climate (see Chap. 3).

The combined effects of population increase, soil degradation, LCC, and climate change pose a great challenge to cope with the population living in drylands (Verstraete et al. 2009). Soil degradation threatens long-term food and biofuel production. Intense irrigation salinizes the land. In Africa and Central America, soil degradation reduces yields on approximately 16% of the agriculturally used land. In sub-Saharan Africa, for instance, cropping land lost at least 20% in productivity over the last 40 years (Sivakumar 2007).

Land can also become unavailable for agricultural production due to oceanic and/or geological processes. Climate-change projections suggest local changes in sea level, wind force, wind direction, and wave energy and/or wave direction. Such changes may lead to LCC by land loss. Land loss due to an increase in wave energy, and/or storm-flood frequency occurs rapidly. Land loss due to sea level rise and/or altered tidal currents, which erode coastal dunes, occurs gradually and over long time. Future challenges are to develop protective and/or coping strategies to avoid land loss from these causes. Policies and decisions have to be made on what, where, and when to apply which measures, and efforts have to be made to optimize their efficiencies. The range of options varies from local, short-term, and effect-focused sea-defense works to developing global strategies to reduce GHG emissions.

Some nations have a long history in coping with local sea-level rise. In the Netherlands, for instance, the lowest city lies 7 m below sea level (SL), and 12.5% and 50% of the land is below SL and below 1 m above SL, respectively. The Dutch developed complicated systems of dykes, dunes, and dams to protect the low-lying land. Such protective systems have to be adapted to future challenges of sea-level rise and land loss.

Landslides in responses to earthquakes or after long-lasting rains cause land loss. Like for flooding landowners are the driving forces to reclaim the land after such natural events. Such reclaiming efforts often lead to LCC and bear the risks of soil erosion and/or loss in biodiversity.

4.4.3 Biofuel

Efforts to mitigate pollution can create political and/or economic pressure to accommodate (more) biofuel production. Many studies argue that substituting gasoline by biofuels is neutral for the annual CO_2 budget (Searchinger et al. 2008). Biofuels sequester carbon through the growth of vegetation. The vegetation takes up CO_2 for photosynthesis. The combustion of the biofuel made of the vegetation releases the CO_2 back into the atmosphere within the next 12 months or so after the vegetation was harvested and distilled into biofuel. To reduce fossil fuel carbon emissions, many states introduced incentives to increase the production and use of biofuel. Consequently, the use of agricultural land changed and new land has been taken into production, both of which mean long-term land-use/cover changes.

It is difficult to assess fully the consequences of LCC due to biofuel production. Uncertainty exists to which extent regional irrigation will change and how efficient plants adapt their water use at high atmospheric CO_2 concentrations. Moreover, the type of LCC and extent depend on economic criteria and subsidence policies. Thus, biofuel production impacts on the atmosphere differ strongly among regions. Studies suggest that if it is economically advantageous to produce biofuel from crops rather than forest, the associated LCC could lead to local and regional cooling at midlatitudes. If the opposite is true, the associated carbon sequestration could cause local and regional warming at midlatitudes, especially during the warm season (Diffenbaugh 2009).

In biofuel production, two mechanisms cause LCC (Kim et al. 2009). "Direct" LCC occur as part of a specific supply chain for a specific biofuel-production facility. The market forces LCC "indirectly." Land that does not belong to a specific biofuel-supply chain is converted. Such changes can include LCC in another part of the world.

In assessing the benefits of replacing fossil fuel by biofuel, the carbon debts have to be considered. This means one has to account for the CO_2 emissions that result from replacing grasslands and/or forests by cropland. Burning of forest to obtain space for biofuel cultures releases CO_2 and causes CO_2 loss from the soil (Reijnders and Huijbregts 2008). Furthermore, the LCC yield net carbon fluxes to the atmosphere for decades due to the LCC themselves. The time needed for biofuels to overcome this carbon debt and to start providing cumulative GHG benefits is called the "payback period."

Various modeling studies assessed the carbon debt and payback periods. The benefit of biofuel and the time to pay back the up-front CO_2 emissions depend on the region, type of LCC, and agricultural management practices (Reijnders and Huijbregts 2008; Kim et al. 2009). Simulations by Searchinger et al. (2008) with a worldwide agricultural model showed that for GHG reductions of 86% excluding LCC, the up-front CO_2 emissions in Brazil pay back in 4 years if sugarcane for ethanol production is grown on former tropical grazing land. If this grassland had been a rainforest before, the payback period lengthened to 45 years. In the USA, biofuels produced from corn on former grasslands increase the GHG emissions by 50%. Over 30 years, corn-based ethanol nearly doubles the GHG emissions. The results by Kim et al. (2009) suggest that the payback period increases with the fraction of forest converted. Using good cropland to expand biofuel cultivation enhances air temperatures similar to converting forest and grasslands for biofuel cultivations. Consequently, using existing good cropland for food production will help to avert GHG from LCC (Searchinger et al. 2008).

Substituting diesel by biofuel reduces the sulfur dioxide (SO_2), but enhances the nitrogen oxide (NO_x) emissions (Reijnders and Huijbregts 2008). The use of biofuel reduces the particulate matter (PM) concentrations emitted and particle size. Smaller particles, however, have more health adverse effects than larger particles (Pope et al. 1995). Furthermore, biofuels increase the soluble organic fraction and oxidation reactivity and influence the nanostructure of diesel soot (Reijnders and Huijbregts 2008).

Most studies focused on CO_2 changes related to biofuel usage. However, any exchange of one fuel by another fuel affects the emissions of other species released during the combustion. The LCC related to biofuel production not only alter the physical states and the CO_2 concentration, but also other atmospheric species. The biogenic emissions of the biofuel cultivations differ from those of the prior land-cover, especially when fertilizers are used. Soybeans or rapeseed, typical crops for biofuel production, emit nitrous oxides (N_2O) (Reijnders and Huijbregts 2008). Microorganisms fix N-compounds from fertilizer and release N_2O in the process. Various studies suggest that N_2O emissions range between 1.25% and 5% of the fixed nitrogen added to a field. Typical fertilizing practice in German rapeseed and Brazilian soybean fields add 1.65 and 1.7 kg of fixed nitrogen per km^2, respectively (Reijnders and Huijbregts 2008). In Brazil, about 45% and 55% of the soybean harvest are for food and biofuel production, respectively. In Germany, about 72.5% of the rapeseed harvest serves for biofuel production.

In addition to the up-front impacts, altered biogenic emissions and different emissions of biofuel, the harvest, refining, and distribution of biofuel affect the atmospheric composition. The practice to burn sugarcane for easier harvest, for instance, releases CH_4 and N_2O into the atmosphere. Corn- or sugarcane-based ethanol corrodes the traditional pipelines (http://www.aaas.org/spp/cstc/briefs/biofuels/) for which it has to be transported by trucks. This transport emits CO_2 that otherwise would not have been emitted. Biofuel burns differently than fossil fuel and may cause technical problems. The technical developments to overcome the problems may have again environmental impacts.

Several other challenges are related to LCC for biofuel cultivations. Experts predict that the large expansion of biofuel cultivations will reduce biodiversity. The monoculture going along with extensive biofuel production may cause soil erosion and increase runoff. The use of fertilizers may increase as some crops suitable for biofuel production need comparatively huge amounts of nutrients. Fertilization may negatively affect water and air quality. As corn production for biofuel expands into drier areas, irrigation becomes necessary. The additional irrigation may cool and moisten the atmosphere locally. The impact of the altered atmospheric composition on atmospheric chemistry and deposition into ecosystem is still a hot research topic.

4.5 Urbanization, Urban Areas, and Megacities

Projections with various GHG-scenarios suggest an increase of the frequency and length of heat waves (Barnett et al. 2005). Such climate change and variability challenge especially the urban population as they worsen the urban heat island (UHI)-related heat stress. A warmer climate in combination with enhanced emissions affects the chemical reaction rates with impacts on air quality. Changes in precipitation associated with a warmer climate affect urban runoff and may challenge the availability of high-quality water.

4.5.1 Urban Heat Island

Many megacities (e.g., Chicago, New York) are located in geographic regions for which climate projections indicate substantial increases of the number of days with temperatures exceeding 32.2 °C or even 37.8 °C.

During summer nights, the UHI effect together with low wind speeds are the main reason for increased heat stress in mid- and low latitudes. The projected increase in ambient temperature together with (increased) UHI effects raise the risk for heat-stress-related health issues and reduce the quality of life (QOL). To mitigate this threat, measures to extend the emergency system and reduce the UHI effect have to be developed.

Parks and tree-covered neighborhoods can create "oasis effects" (Taha et al. 1991). Urban forests and parks reduce not only hydrocarbon concentrations, but also the UHI effect directly by shading the ground surface and indirectly by evapotranspiration. On average, air temperature decreases about 1 K per 10% canopy cover. Thus, neighborhoods with trees are about 2–4 K cooler than those without trees (e.g., Fig. 4.11). In Davis, California, for instance, air temperatures in an isolated orchard were 4.5 K lower than in the urban upwind (Taha et al. 1991).

Some cities already started efforts to reduce heat-stress due to UHI effect, urban growth, and climate change. The City of Chicago, for instance, plans to add approximately one million trees to their 4.1 million trees by 2020. Other cooling and energy-efficiency efforts persuaded include cool roofs, parks, and green roofs.

Studies suggest that white roofs can decrease urban daily maximum and minimum temperature by 0.6 and 0.3 K, respectively. Simulations using an urban canyon model coupled to a global climate model suggest that on average over all urban areas globally, installing white roofs may decrease the annual mean UHI effect by 33% (Oleson et al. 2010). The local effect of white roots depends on the season, location, and size of the urban area. At high latitudes in winter, for instance, an increased roof albedo affects marginally the UHI effect because roofs are snow-covered and incoming solar radiation is low. In low latitudes and midlatitude summers, white roofs help diminish the UHI effect and reduce emissions, as air conditioning needs less energy. In midlatitudes white roofs may lead to increased emissions from heating in winter.

Parking lots are sources of motor vehicle pollutants and have the character of small heat islands. In many European cities, homeowners installed "green parking lots" for their second vehicle. Green parking lots like green alleys are made of porous pavement often with about 10 cm in diameter small grass patches. These surfaces permit water to infiltrate into the soil instead of going into the sewer system. Due to evaporative cooling from the grass and open soil patches, green parking lots reduce the heat buildup.

Parks, urban forests, green parking lots and green roofs may require irrigation. Daytime Bowen ratios are related inversely to the irrigated area (Grimmond and Oke 1995). Thus, enhancing irrigation in urban

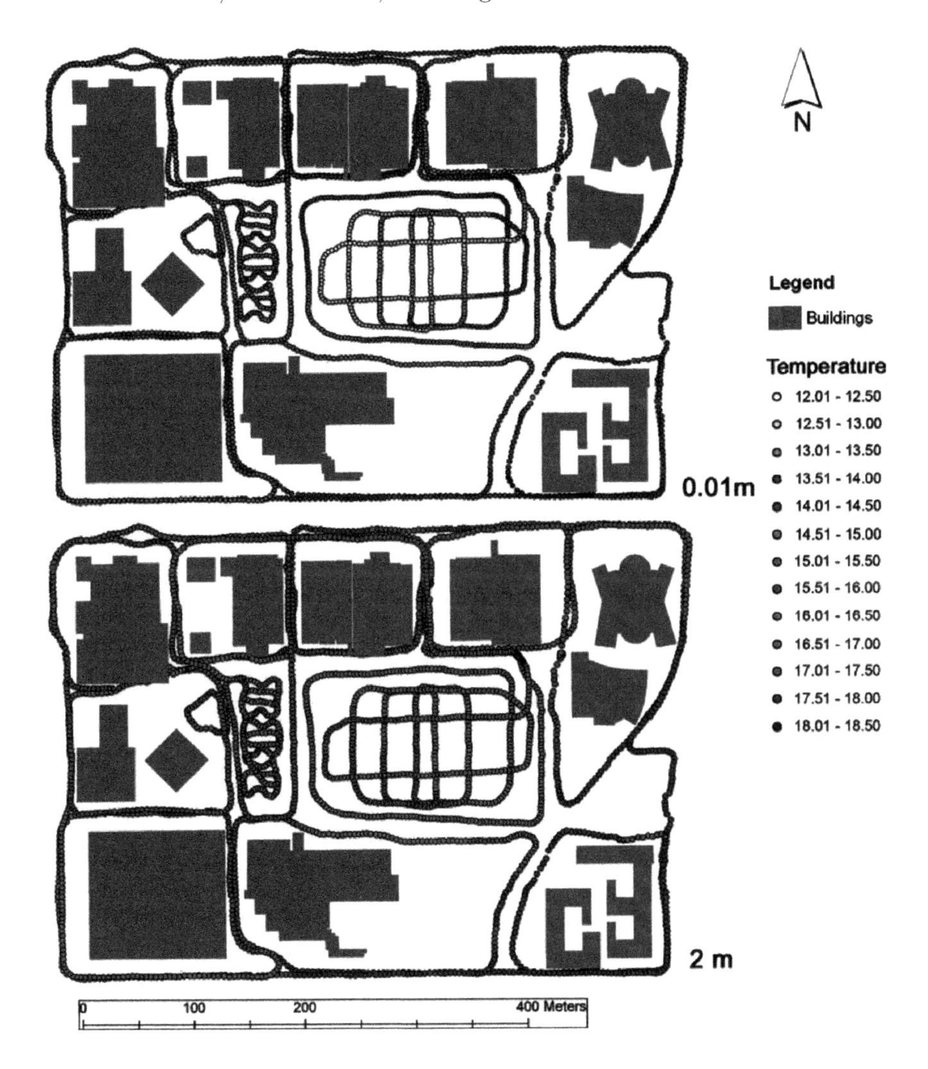

Figure 4.11. Temperatures from measurement drives from 0530–0630 LT as observed at 0.01 (*top*) and 2 m height (*bottom*) (From: Chow et al. 2011)

areas may reduce the UHI effect. Parks, urban forest, white roofs, green parking lots and green roofs mean LCC within the city and introduce heterogeneity on the local scale. In addition to reducing temperature and enhancing evapotranspiration, these measures may affect precipitation via secondary processes and have implied changes in stormwater management.

4.5.2 Water Management

Some climate projections suggest intensifying of extreme weather events (Huntington 2006). Increased precipitation rates during thunderstorms may provide more water to the sewer systems than they were designed for. As the cities grow, the sealed area expands. To prevent flooding, diseases, and maintain water quality, cities have to develop stormwater-management plans (Hunt and Watkiss 2011). Green roofs, alleys and parking lots, parks, and urban forest can be part of such management plans.

In high latitudes, more solid precipitation in the lee of cities may increase the likelihood for roof collapses and the need for clearing the roof and/or the need for new building codes. Changes in groundwater levels in response to the altered precipitation distribution may affect households on the outskirts of high-latitude cities as they mostly rely on private dwells.

4.5.3 Air Quality

In the last decade of the twentieth century and the first decade of the twenty-first century, total anthropogenic emissions have decreased in western countries. Anthropogenic emissions increased from international shipping due to globalization and in the growing urban centers of developing countries. Of the worldwide emissions, about 75% occur in urban areas. As more people live in megacities, urban air pollution may affect the health of a large fraction of the world's population adversely. Ozone, for instance, which forms in polluted cities with high insolation, has strong health adverse effects in young children and elderly.

Urban air pollution is a direct or indirect consequence of fuel consumption for transport, industry, or domestic use and on the raise. The air pollution problem is especially acute in developing or strongly industrializing countries where cities with population $\geqslant 10^6$ grow rapidly (Molina and Molina 2004). In India, for instance, 300 million people live in megacities. In megacities, about 70% of the emissions stem from energy production for heating and cooling. About 20% are due to traffic. Concurrent with the growth of cities, the number of vehicles went up by a factor of four since the 1950s. Traffic strongly contributes to the NO_x and PM emissions and provides ozone precursors. Some pollutants stem from food-preparation habits. For instance, grilling or barbecuing of meat and smoking of fish involve combustion like burning of fossil fuel or biomass and produce particulate polycyclic aromatic hydrocarbons (PAH) that may be carcinogenic, mutagenic, and teratogenic. As megacities grow, the concentrations of these pollutants increase as well (Molina and Molina 2004).

In some megacities, measurements of meteorological and chemical conditions were performed. Chemical studies most often focused on a compound that currently causes an air pollution problem. Lee et al. (2011), for instance, measured the ambient particulate PAH at Seoul, Korea. Over the 18-month measurement period, PAH concentrations ranged between 1.57 and 166 ng/m^3 with 26.6 ± 28.4 ng/m^3 on average. The increased fuel consumption in Seoul and in Northeast Asia in winter explains the winter maximum and summer minimum. In Seoul, ambient PAH concentrations were comparably lower than those found in cities in China, but exceeded those in other cities. Since the formation of PAH is temperature dependent, the relationship between the ambient particulate PAH concentrations and the inverse of the ambient temperature ($1/T$) can be used to assess whether PAH is locally formed or stems from advection. Sites with close by emissions typically have stronger temperature–concentrations relationships than background sites. Comparison of the Seoul PAH-to-temperature ratio with that of a background site (Gosan) indicates that in Seoul, advection from outside Seoul and local emissions dominate the PAH concentrations (Lee et al. 2011).

The increasing size and number of megacities requires strategies to keep the pollution response to urban growth low. Ozone-mitigation programs are required that reduce emissions from traffic and power plants. Some European countries restrict driving when the ozone concentrations exceed a critical value. In American megacities, extension of public transit systems and bike-path systems are measures considered frequently for reducing traffic emissions. In Chinese megacities, electric motorbikes outsource the emissions. The emissions to produce the power needed to operate the motorbikes occur elsewhere. This procedure means an energy loss. The chemical energy of the fuel is converted first to electrical and then to mechanical energy.

The deteriorating urban air quality affects the natural and agricultural ecosystems near megacities. Airborne pollutants from major conurbations influence air quality, weather, and climate not only on the local (in town) and regional scale (in the immediate downwind), but also on a continental scale. Consequently, megacities and urbanization with their associated concentrated emissions and later advection of aged polluted air may become an international challenge.

Increasing urbanization poses various questions for the assessment of future atmospheric composition and climate. Questions to be addressed are how to develop storylines for urban emissions; how to use estimates of atmospheric emissions, composition, and related processes in assessments of long-range chemical transport; the vulnerability of ecosystems due to urban growth, and what are the impacts of urbanization on weather and climate. Uncertainty lies, among other things, in the exchange of biogenic to anthropogenic emission rates and emitted species, the locations where urbanization may occur, and the timeline.

Studies on future air quality focus usually on a region that already has a major population center. These studies, to assume reasonable LCC, exclude geographic barriers (mountains, lakes, rivers) and politically difficult or expensive to change areas (e.g., forests, drainage of marshland) from urban growth, and favor easy to change areas (e.g., grassland, cropland) as areas of future urban growth. Assumed emissions are related to expected economic development and are orientated at current major economic sectors. Typically, focus is on a pollutant or group of pollutants with a history of causing exceedances of the current air quality standards, that are adverse to health, and/or that are expected to occur at higher frequency and/or concentrations in a warmer climate and/or a more densely populated area. The increase in ozone concentration in urban areas under warmer climate conditions, for instance, poses a threat to human health and to ecosystems and agriculture in the downwind of conurbations.

Since computational time is high for photochemical modeling, many studies on air quality in growing urban areas focus on an episode in a base year and an assumed future year. Civerolo et al. (2007), for instance, examined the impact of urbanization between 1990 and 2050 on O_3 concentrations. They performed 18d-simulations with a regional air quality forecast (AQF) model package with same emissions, but different land-cover distribution. The assumed 2050 land-cover for the New York metropolitan area followed the storyline of the A2-scenario. The afternoon near-surface air temperatures and atmospheric boundary layer (ABL) heights increase by more than 0.6 K, and 150 m, respectively, while water-vapor mixing ratios decrease by more than 0.6 g/kg where the urban area increased. The impacts of the urbanization on O_3 concentrations are more complex (Fig. 4.12). The spatial patterns of O_3 changes are heterogeneous with decreasing ozone concentrations in some areas. Increasing the anthropogenic emissions to be consistent with the urban growth increased the O_3 concentrations in outer counties of the metropolitan area. Ozone concentrations decreased along the coasts of Connecticut and the Long Island Sound among other places.

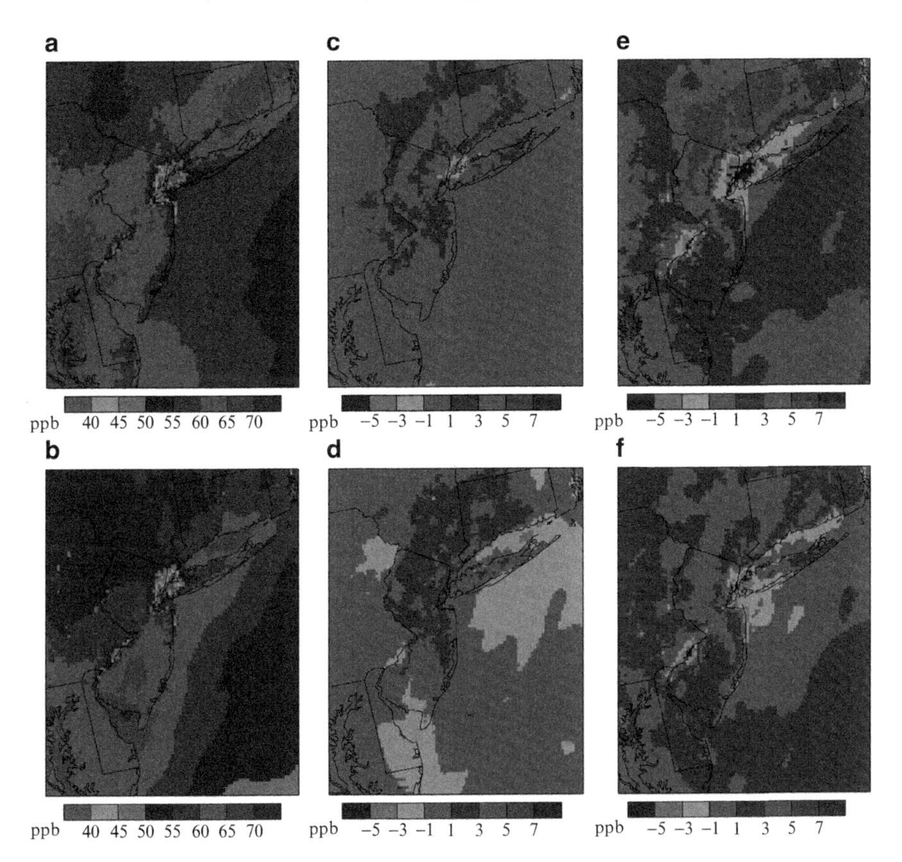

Figure 4.12. Average surface O_3 concentrations for the (**a**) reference 1993 simulation, (**b**) the reference 2056 simulation and differences between the future model simulations and the respective reference simulations (future-reference) for (**c**) 1993 episode with 2050 A2 land-cover and 1990s emissions and (**d**) 2056 episode with 2050 A2 land-cover and 1990s emissions, (**e**) 1993 episode with 2050 A2 land-cover and 2050 A2 emissions, and (**f**) 2056 episode with 2050 A2 land-cover and 2050 A2 emissions (From Civerolo et al. 2007)

The assumed urbanization increases the episode-maximum 8-h O_3 levels by more than 6 ppb and the episode-average O_3 levels by about 1–5 ppb over the metropolitan area.

In urban areas, wettable and soluble particles may be emitted, and secondary aerosols may form in the polluted air. Urbanization seals large parts of the formerly open land. This means that less frequently soil and other particles (e.g., pollen, yeasts, particles from plant decay) become airborne. Thus, urbanization alters the spectrum and availability of CCNs and/or INs. Changes in CCN/IN in response to altered emissions associated with urbanization and/or political changes were found to affect precipitation. Bokwa (2010), for instance, examined the relationship between the decrease in emissions after Perestroika and precipitation. The time before Perestroika (1972–1988) was characterized by high and, the time thereafter (1989–2005) by, relatively lower emissions. The authors used daily precipitation data from 1971–2005 from one urban site in and three rural sites around Kraków, Poland. No trend in the precipitation totals were detected for 1971–2005. The precipitation totals of 1972–1988 differ significantly from those of 1989–2005 for only one of the rural sites in winter and for the Kraków site in fall. The improved air quality affected the spatial precipitation pattern mainly in spring and fall for daily-accumulated precipitation greater than 5 mm.

4.5.4 Urban Planning and Development

Urban planning and development affect where and which kind of LCC due to urbanization occur. The future may bring two different types of urbanization. Economic reasons may require more of the work force to live in (mega)cities. An aging population and increased demand for recreational zones along the coasts will lead to LCC for housing and recreational facilities (e.g., golf courses).

Recreational zones typically have urban forest, parks, and other green areas to keep a pleasant microclimate. Thus, the UHI effect remains low in recreational areas even in warm climate zones. The recreational urban areas may become local oases with relatively cooler and wetter conditions than the surroundings, given enough water is available to sustain the urban vegetation. In warm climates, emissions from domestic heating are rather low; traffic and generation of power for electricity and air conditioning are the major anthropogenic emission sources.

For many megacities in developing countries, no or hardly any planning related to the LCC exists. Megacities face environmental problems related to the UHI effect, air quality, and ventilation. Ventilation refers to local and regional wind systems handling the exchange of fresh and polluted air. Wind and subsequent ventilation conditions govern the climate and air quality of any city. For cities in complex terrain and/or at large lakes or the ocean, mesoscale circulations may be relevant for the ventilation and city's climate (Fehrenbach et al. 2001). The knowledge of local and/or mesoscale circulation systems can be used in local planning to improve the QOL and for hazard assessments. If the local wind system, for instance, may become very strong, it may pose the risk of hazards for high buildings in the affected area. The strength of local wind systems determines the intensity of the UHI effect. Ventilation paths into cities may advect clean or polluted air especially during nights with cold air production in the adjacent rural area. In mid- and low-latitude cities, advection of cold air reduces the UHI effect. In high-latitude cities located in valleys, the drainage of cold air may enhance inversion strength.

In some (growing) cities, environmental problems become particularly bad due to direct and indirect couplings of air pollution, UHI effects, and local or mesoscale circulation systems (Oke 1995). In Barcelona, Spain, for instance, O_3 concentration can become very high as the surrounding mountains and the sea breeze trap the pollution in the city. The nocturnal land breeze advects pollutants onto the Mediterranean Sea. The next day's sea breeze then advects aged polluted air. Over time, daytime pollution levels raise.

The paths for clean air and/or cool air advection have to be identified for growing (mega)cities. Authorities have to ensure that no major emission sources are inserted into these ventilation paths. If emission sources exit in the ventilation paths, emission-reduction measures have to be taken. Growing cities with reduced ventilation, high emissions, and frequent temperature inversions are extremely receptive to air pollution. In these cities, emissions should be reduced as far as possible.

Structural properties like aerodynamic roughness length, zero-point displacement, effective building heights, and density of the urban canopy affect local atmospheric conditions. New constructions in ventilation paths have to be avoided when they increase surface roughness and/or act as flow obstacles. Flow obstacles and high surface roughness weaken nocturnal drainage flows. If ventilation paths are in areas with dense building structures, measures may be necessary to improve ventilation.

Since city centers, parks, and suburban, commercial, and industrial areas interact differently with the ABL, local climates differ among these parts of a city. The natural and anthropogenic emissions differ among these parts of a city as well. Thus, ensuring a mix of different urban surfaces may reduce the UHI effect that otherwise would occur locally in response to the growing (mega)city.

The climate of growing megacities depends strongly on intelligent planning or the lack thereof. Any urban planning requires the analysis of huge amounts of data from various sources (e.g., land-use, land-cover, emission, topography, soil, and meteorological data) and knowledge of local relationships (Hunt and Watkiss 2011). Planning criteria have to be identified and/or designed. The huge amount of data and relationships may make the planning cumbersome. Some of the data interpretation may also be nonobjective. Thus, decisions may be irreproducible for similar cases when officers change.

Fehrenbach et al. (2001) suggested an automatic tool that provides a knowledge-based classification for planning objectives. The conceptual model relies on spatially distributed data of land-cover and topography from a Geographic Information System (GIS). The model uses conceptual relationships concerning land-cover and ventilation, the dominant controls of local climate and air quality and a knowledge-based classification scheme for climate analysis to identify automatically critical areas with respect to ventilation, air quality, and thermal effects. The model uses three different information layers to classify each $100 \times 100\,\mathrm{m}^2$ grid element. It assigns zero or one planning objective per problem section to each grid element. Critical areas typically have a combination of three different planning objectives. Uncritical areas are not associated with any objective. The model provides reproducible maps of urban climate and air quality that can be evaluated by meteorological and air chemistry measurements.

This conceptual model is objective and provides a high degree of transparency. Therefore, maps created by the model are well suited for urban and regional planners in their initial planning. These maps cannot replace the legally required environmental impact assessment studies. However, they make preparing and designing the network for the meteorological and air chemistry measurements and identifying cases for regulatory photochemical modeling more cost efficient (Fehrenbach et al. 2001).

4.6 Detecting Land-Cover Changes in Observations

As discussed in Chap. 3, impacts of LCC on local climate are difficult to detect in measurements. Typically, long-term monitoring data are taken over extended patches of grass or bare soil. Thus, the data rather represent the advected response to upwind LCC than a change at the site. Association of a change in the microclimate at a site requires additional knowledge. It has to be excluded that relocation of the site caused the change. Changes in site location mean a different latitude and/or longitude and potentially different elevation and exposure to dominant major flow patterns (e.g., drainage flows, mesoscale circulations, emission sources). These changes may affect the measured air and dewpoint temperature, their minimum and maximum values, wind speed and direction, and concentrations. Consequently, even if the new and old locations have the same land-cover type, the sampled data may change due to the relocation.

Other difficulties with long-term measurements are observers' preference to report precipitation as values divisible by 5 and/or 10 and/or underreporting. Such behavior may induce trends in long-term data when observers change. Long time series may have errors from digitizing old paper records, and the procedure of filling missing data. Such uncertainty makes it difficult to correlate observed changes in local climate with LCC in a site's surroundings.

Sampling of long time series takes decades. Therefore, scientists have to assess the suitability of data from existing long-term monitoring networks for their scientific purposes. Before using long-term data for detection of LCC impacts, scientists have to assess the data's uncertainty and limitation, and availability of urgently needed data. The use may require developing intelligent analysis methods to ensure meaningful conclusions.

The recent decline of the worldwide gauging station network makes it even more difficult to detect the impacts of LCC on regional climate. PaiMazumder and Mölders (2009) investigated how network density and design may affect regional averages. They created a gridded dataset with $50 \times 50\,\mathrm{km}^2$ increment over Russia for July and December 2005, 2006, and 2007 using the Weather Research and Forecast (WRF) model. By using this data, they determined regional averages of wind speed, temperature, and relative humidity for $2.8^\circ \times 2.8^\circ$ areas as the "reference" or "truth." They designed 40 artificial networks on the model domain that encompassed ten networks of 500, 400, 200, or 100 different randomly taken WRF-grid-points as "sites." In addition, to these 40 networks, they projected the location of the 411 sites of an existing network – called "real" network hereafter – onto the model domain. The design of the real network was biased by site accessibility and preference for agriculturally used land. By using the WRF-simulated values at the "sites" of the 41 networks, regional averages were calculated and compared with the "reference" regional averages determined using all WRF-simulated values. Their results indicated that low-density networks (e.g., the ten 100-sites-networks over Russia) have difficulty to represent the regional distribution of sea-level pressure. Regional averages calculated based on the 30 randomly distributed networks with 200 or more randomly distributed sites represent the reference regional averages and temporal and spatial variability for all quantities well. Summer regional precipitation averages are very sensitive to network density and distribution. High-density randomly distributed networks can better capture strong convective precipitation events than coarse networks (PaiMazumder and

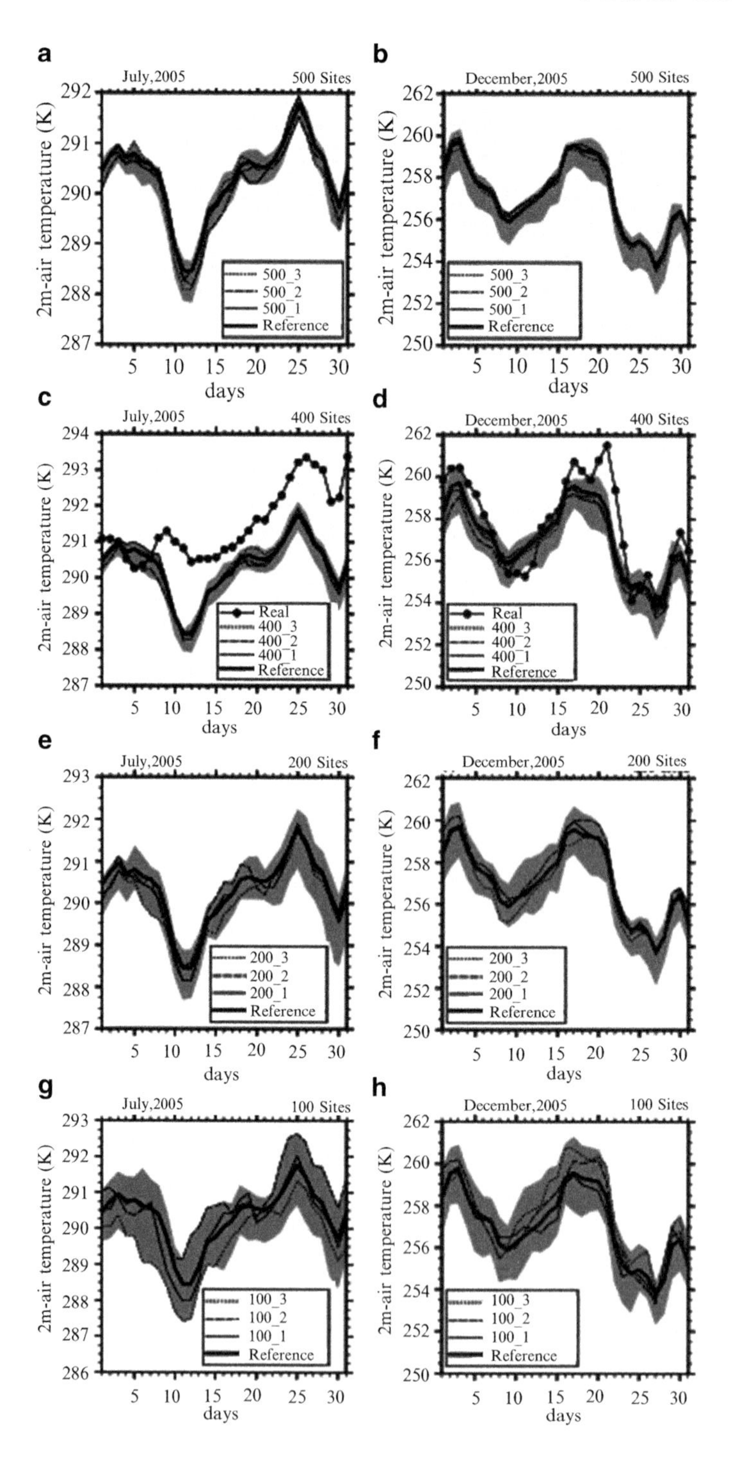
a
July,2005
500 Sites
2m-air temperature (K)
days
500_3
500_2
500_1
Reference
b
December,2005
500 Sites
c
July,2005
400 Sites
Real
400_3
400_2
400_1
Reference
d
December,2005
400 Sites
e
July,2005
200 Sites
200_3
200_2
200_1
Reference
f
December,2005
200 Sites
g
July,2005
100 Sites
100_3
100_2
100_1
Reference
h
December,2005
100 Sites

Mölders 2009). This means that randomly distributed high-density networks may be needed to identify impacts of LCC.

The authors reported that randomly distributed networks represent the regional averages better than "biased" networks, even if the latter have more sites (Fig. 4.13). Regional averages derived from the real network differed greatly from the reference regional averages. The misrepresentation of the landscape by the real network caused errors in geographical and temporal trends for most quantities. The real network represented the regional averages best over flatland; discrepancies in regional averages were lower over shores of large lakes than coasts and increased with terrain height and complexity. Temporal offsets in regional averages existed for frontal passages when several sites of the real network were passed at nearly the same time. Networks with randomly distributed sites show no such geographical trends (PaiMazumder and Mölders 2009).

Typically, the real network overestimated the regional averages of 2 m temperature, soil-temperature, and downward shortwave radiation up to 1.9 K, 1.5 K, and 19 W m^{-2} and underestimated the regional averages of sea-level pressure, wind speed, and precipitation up to 4.8 hPa, 0.7 m/s, and 0.2 mm/d in July, respectively. In December, the respective overestimation amounted 1.4 K, 1.8 K, and 14 W m^{-2}, while the respective underestimation of the regional averages was 4.8 hPa, 0.5 m/s, and 0.5 mm/d (PaiMazumder and Mölders 2009). These values are of similar magnitude than changes due to LCC. This means that LCC impacts on regional climate cannot be identified easily in observations unless they LCC have large extent and the network has high density and random distribution. Moroever, shutting down of sites may alter the regional average temperatures calculated from the lower number of sites (Fig. 4.13),

Figure 4.13. Time series of regional averages of 2 m air temperature as obtained for the reference data using all WRF data within all 2.8° × 2.8° areas and as derived for various networks based on the "sites" within the 2.8° × 2.8° areas for selected 500-sites-networks in (**a**) July 2005, (**b**) December 2005; 400-sites-networks in (**c**) July, (**d**) December; 200-sites-networks in (**e**) July, (**f**) December; and for selected 100-sites-networks in (**g**) July and (**h**) December. The solid line with filled circles in (**c**) and (**d**) represents the regional averages derived from the grid-cell values at the points of the real network (411 sites). All other lines represent the regional averages with lowest error values among the ten setups of the respective network; the shaded area represents the maximum over- and underestimation of the reference regional averages that occurred for the ten networks of same density (After PaiMazumder and Mölders 2009) that is, older regional averages have to be recalculated using only the remaining sites for comparison with regional average after the shutdown of sites.

As pointed out by PaiMazumder and Mölders (2009), the purpose of the real network was to collect soil-temperature data for agricultural purposes. Therefore, it represents the conditions for agriculturally used land, but not the landscape for which the regional averages were to be determined. Surface fluxes, temperature, and humidity as well as development of convection over cropland differ strongly from those over forest. Thus, a network biased with respect to one land-cover type will provide incorrect regional averages for the variables of state and fluxes. The great importance of the representativeness of a network was also found using other methods. Pielke et al. (2007) found that near-surface temperatures sampled at poorly and inhomogeneously sited stations vary stronger as compared to North American Regional Reanalysis data than well-sited stations do.

Despite being superior from a theoretical point of view, networks with randomly distributed sites are hard to establish for various reasons. In densely populated areas, there may be property issues or issues with the fetch conditions. In remote areas and mountainous terrain, the maintenance can be extremely expensive, especially when measurements are to be made over several decades. In cases like the above-discussed example of determining regional averages from nonrandom networks, for instance, methods to remove the geographical/temporal trends have to be developed.

Keeping all these points in mind, the following conclusions can be drawn. The impacts of LCC on regional climate can more likely be detected by means of a long-term monitoring network under the following three conditions: The fractional cover of one land-cover type increases notably, a new land-cover type is introduced at large extent (increase of diversity), and/or a land-cover type vanishes (loss of diversity).

A major problem with LCC-induced nonclassical mesoscale circulations (NCMC) is that conventional meteorological networks do not allow observing them. These circulations often generate shallow cumuli that are difficult to detect in satellite images due to their small size and short life span.

4.7 Changes of Snow and Ice

4.7.1 Snow-Vegetation and Permafrost-Vegetation Relation

Permafrost denotes soils that remain at temperatures below the freezing point for two consecutive years. In this sense, permafrost is not directly a land-cover. However, permafrost and the active layer depth (thickness of the upper soil layer that thaws in summer) determine which type of vegetation can survive on permafrost grounds.

Permafrost exists in high latitudes and at high elevation. Freeze-thaw cycles and the related release of latent heat and consumption of energy influence the thermal and hydrological properties of permafrost soils. The thermal conductivity of ice, for instance, exceeds that of water about four times. Frozen ground hinders infiltration, capillary action, and percolation of water. The presence of soil ice alters soil thermal fluxes through the dependence of soil thermal conductivity and volumetric heat capacity on volumetric soil water and ice content.

During high-latitude winter and at high elevation, air temperatures are low and the atmospheric surface layer (ASL) experiences frequently stable stratification. Evaporation, sublimation, and transpiration are marginal compared to moderate weather conditions. Thus, the frozen soils retain their moisture. During summer, the active layer thaws. The soil water becomes available for evapotranspiration. In fall evapotranspiration decreases and rain refills the active layer to saturation before freeze-up.

As climate changes, the extension of continuous and discontinuous permafrost as well as active layer depth and the frequency of soil frost in midlatitudes during winter change. A deeper active layer may introduce LCC as vegetation (e.g., deciduous forest) that needs deeper active layers, may migrate into these areas. The increase in active layer depth also means an increase of the plant available water and potential water supply to the atmosphere. The increased atmospheric water supply affects the stability of the ABL, cloud, and precipitation formation.

The vegetation changes themselves affect the permafrost and active layer depth with further consequences for active layer depth, and the distribution of permafrost and vegetation or for restoration of conditions. The annual cycle of temperatures for soils covered by different vegetation differs strongly (e.g., Fig. 4.14) because different vegetation types shield the ground differently and take up water from different levels. As the active layer depth increases, vegetation with higher water demands can migrate into these areas.

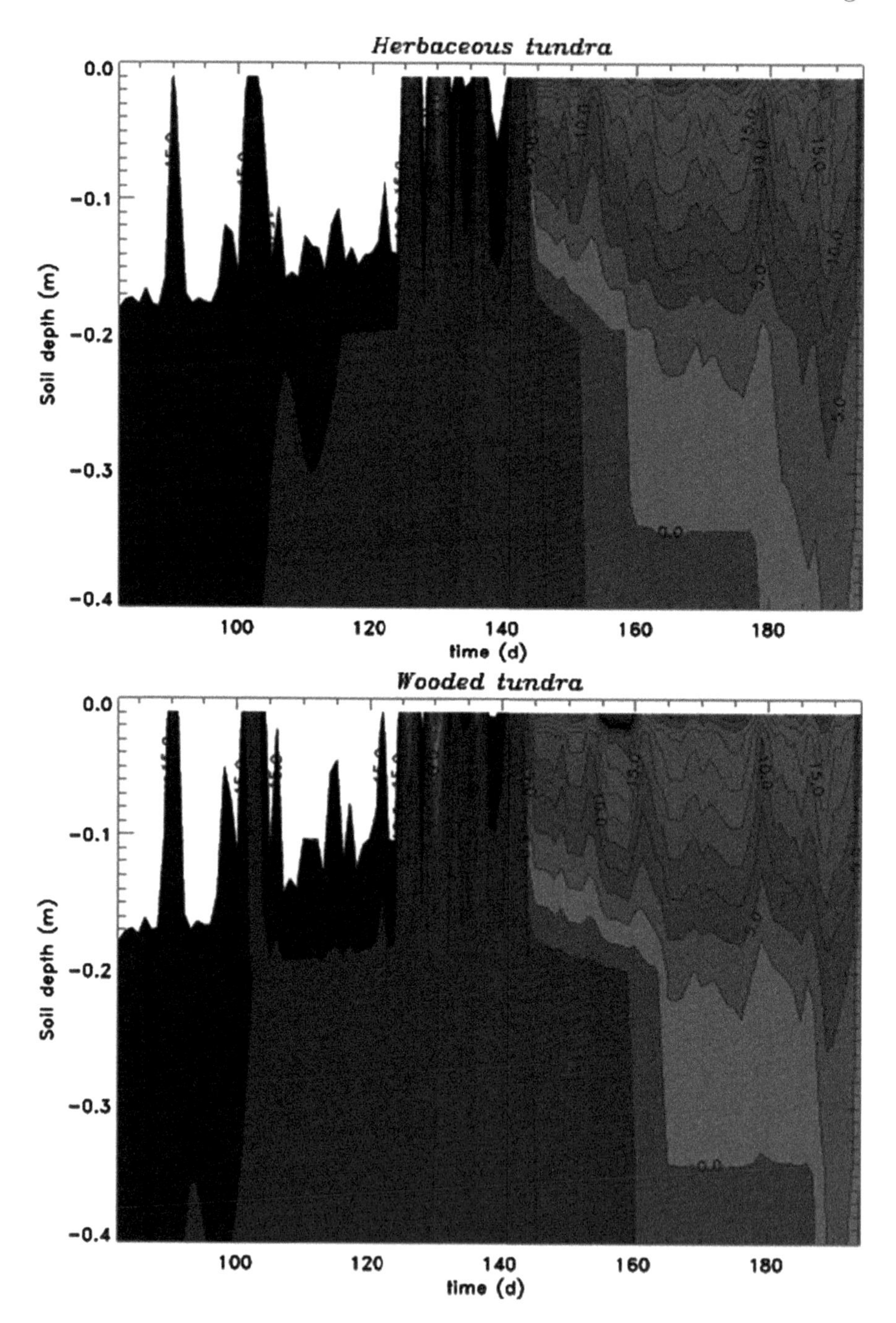

Figure 4.14. Comparison of soil temperatures for soil covered by herbaceous tundra (*top*) and wooded tundra (*bottom*) as simulated over one warm season with the hydro-thermodynamic soil vegetation scheme (HTSVS). The simulations were forced with the same meteorological observations

Restoration of permafrost may occur via the following mechanism. As more deciduous trees grow in an area, they produce substantial shadow during the summer, and reduce the irradiation. The soil remains cooler compared to less dense forest conditions and tundra-covered or bare soil. Conifers prefer cool soils and migrate into these areas. Once large enough, they provide substantial shadow year round. The large interception storage of conifers leads to a less thick snow-cover underneath. Thus, for the same atmospheric, soil, and geographic conditions the soil cools stronger under coniferous trees than soil under deciduous trees during the cold season. In spring, the cooler active layer requires more energy to thaw. This additional energy is not available, for which the active layer depth is reduced further. The reduced active layer depth worsens the survival conditions for deciduous trees. If these effects offset the warming due to climate change, over time land-cover may change from deciduous-forest dominated to coniferous forest.

Changes in permafrost distribution affect the atmospheric conditions via LCC and GHG by biogeochemical and biogeophysical feedbacks. As permafrost thaws, it releases water vapor, CO_2, and CH_4 among other gases. The released GHG may lead to further warming and thawing of permafrost. The new land-cover that may grow in areas of thawed permafrost, or permafrost with increased active layer depth, may have different biogenic emissions.

In many regions, snow covers the ground at least seasonally. It insulates the underlying soil and reduces or suppresses the heat exchange between the soil and atmosphere. Changes in snow cover and snow-cover days affect the energy budget due to the differences in albedo and emissivity of snow-covered and the snow-free land surfaces. The albedo of a dark soil, for instance, is about 0.1, while that of fresh snow is about 0.8; emissivity decreases from 0.82 for old snow to about 0.95 for dark soil (Pielke 2002). An increase/decrease in albedo feeds back to a decrease/increase in temperature (temperature-albedo feedback). Changes in solid precipitation amount affect frost-depth with implications for the onset of green-up, and survival of ecosystems.

If climate change alters the location of areas with frequently broken snow-cover, the local wind and temperature climatology will change. The sensible heat fluxes of snow-free areas within an otherwise snow covered area exceed substantially those of the snow that surrounds them. The increased sensible heat fluxes may enhance vertical mixing in the ASL and induce convection. Under favorable large-scale conditions, the spatial differences in surface temperatures may produce NCMC.

4.7.2 Ice–Vegetation Relation

Sea-ice is not a land-cover, but an ocean cover. However, there are indications that sea-ice distribution and extent may influence land-cover. A future challenge is to assess the impacts of the changing sea-ice extent on land-cover and climate. As climate warms, larger fractions of the polar waters may be open for a longer time allowing ship traffic that will affect air quality. In sea-ice, areas of open water (polynyas) may occur more frequently or with larger extent. Continental ice-shelves may collapse. It is still an open question whether shelf-ice collapses/retreats and polynyas affect vegetation via locally altered atmospheric conditions in a similar way as an altered sea-ice distribution does. Nevertheless, the mechanisms by which polynyas and ice-shelves affect the atmosphere are reviewed for completeness.

4.7.2.1 Sea-Ice

GCM simulations with various fixed sea-ice distributions indicate that moderate and large sea-ice anomalies enhance temperatures over the adjacent land (Bhatt et al. 2010). Regional model simulations with and without consideration of sea-ice in the Bering and Chukchi Seas showed that the absence of sea-ice increases surface temperatures by 10–15 K (Walsh et al. 1993). The consequently warmer ABL modifies sea-level pressure, near-surface wind, and precipitation. The altered static stability in the ABL affects the surface wind stress and the exchange of momentum at the surface. These and other evidence, though not conclusively, suggest that sea-ice decline may drive LCC on near-coastal land.

Sea-ice affects the atmosphere via the energy and water budget. Sea-ice albedo values exhibit a broad range depending upon surface cover. Values of about 0.9 are typical for snow-covered ice, but can reach up to 0.97 for fresh snow. The average albedo of first-year snow-covered sea-ice and young gray ice without and with snow are about 0.74, 0.2, and 0.3, respectively (Grenfell and Perovich 2004). The albedo of sea-ice topped with melting snow is less than of snow-covered sea-ice. In mid-summer, albedo ranges from 0.1 for deep dark ponds on ice to 0.65 for bare white ice. Open water, except for the sun glint, has the lowest albedo ($\approx$0.05). Thus, net radiation and surface-energy budgets differ strongly between sea-ice and the ocean with consequences for near-surface temperature. The sensible heat fluxes over open water exceed those over sea-ice notably. The enhanced water-vapor fluxes over open ocean may increase the atmospheric water-vapor content. All these effects modify near-surface temperatures.

Ecosystems are temperature and water sensitive. Indeed, observations document rapid vegetation changes along the Lewis Glacier and Baffin Island, Canada. Bhatt et al. (2010) used Normalized Difference Vegetation Index (NDVI) and high-resolution satellite-derived sea-ice data for 1982–2008 and found a coherent temporal relation between near-coastal sea-ice, land-surface temperatures, and NDVI. During 1982–2008, early-summer ice breakup within 50 km of the coast declined 25%, on average, for the entire Arctic and up to 44% in the East Siberia to Chukchi Seas. These changes in sea-ice distribution affected not only the ecosystems along the coast, but also far inland due to temperature changes. In low-elevation (< 300 m) tundra, the summer warmth index (SWI), as defined by the sum of the monthly mean temperatures above freezing, increased 10–12 K/month on land along the coasts of the Chukchi and Bering Seas and on average 5 K/month (24%) in the Pan Arctic. In North America and Eurasia, the SWI on land increased 30% and 16%, respectively. Percentage-wise, the land warming was stronger ($> 70\%$) close to the Greenland Sea, Baffin Bay, and Davis Strait. Concurrently, the NDVI increased across most of the Pan Arctic except over land regions along the Bering Sea and West Chukchi Sea coasts. Percentage-wise, the NDVI increased greatest (10–15%) over the land in the North America High Arctic and along the Beaufort Sea. The land along the Beaufort Sea also experienced the strongest absolute increase in maximum NDVI. The springtime sea-ice concentrations and increases in SWI and NDVI correlate consistently. This correlation indicates a link between sea-ice decreases, warmer land temperatures, and increased vegetation productivity (Bhatt et al. 2010).

4.7.2.2 Polynyas

Polynyas are surface-cover changes that to a certain degree are influenced by the atmosphere, but also influence the atmosphere. In past decades, some polynyas remained open for multiple winters (e.g., Weddell Polynya 1974–1976). Two types of polynyas exist that differ by their mechanism of formation. Latent heat-flux polynyas are the open water area between the coast and first year pack ice. This type of polynya builds in first year pack ice that was pushed seaward by ocean currents, cyclones or katabatic winds, fast ice, or an ice bridge. In the open water, new ice forms that again is pushed toward the first year pack ice increasing the amount of first year pack ice. Sensible heat-flux polynyas are driven thermodynamically. They form where the water-surface temperature stays above the freezing point of ocean water (271.2 K) due to upwelling warm water.

In both cases, polynyas provide huge amounts of sensible and latent heat to the atmosphere as compared to the adjacent ice-covered ocean. The huge latent heat fluxes can create fog. Another important consequence of the strong heat fluxes from the open ocean is enhanced sea-ice formation and subsequent salt rejection to the underlying water column.

The thin new ice (frazil ice) that builds in polynyas dampens waves for which the polynya is smoother than the rough pack ice. The strong heat fluxes also warm the air that moves over the polynya. The increase in temperature depends on the latent heat fluxes that are a function of the environmental conditions (e.g., wind speed, relative humidity) and the extent of the polynya. Typically, polynyas are wider for high than low wind speeds.

As cold katabatic winds reach the polynya, turbulence is created. The exchange of the ASL and the overlying atmosphere increases and the downward transfer of momentum from the core of the katabatic airflow increases.

The relatively large horizontal temperature gradient between the air over the ice-sheet and the polynya can cause an ice-breeze that enhances the katabatic wind speed. The intensity of the ice-breeze depends on the width of the polynya. The subsidence of maritime air over the ice-sheet can strengthen the inversion and increase the negative buoyancy thereby accelerating the katabatic wind in the coastal region.

Over large polynyas, neutral layers of up to 500 m in height may form. The pressure-gradient force in this layer acts in the same direction as the katabatic force. Consequently, the flow accelerates in the offshore direction.

At a distance of the coastline that corresponds to the Rossby radius of deformation, the offshore component of the wind vector weakens. The flow becomes more along shore. This change in wind direction affects the pack-ice distribution as offshore wind pushes ice away from the coast.

4.7.2.3 Shelf-Ice

Along the coasts of Antarctica and the Arctic Ocean, ice-shelves exist. In a warmer climate, the extent of ice-shelves decreases. At the location of an ice-shelf collapse/retreat, the radiative, aerodynamic, thermal, and hydrological conditions at the surface–atmosphere interface change drastically, with feedbacks to weather and climate (Fig. 4.15).

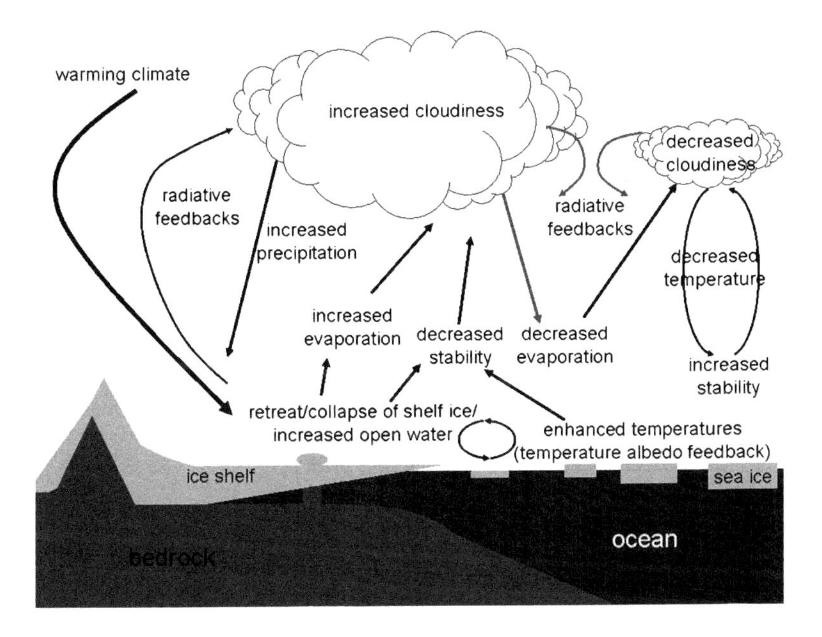

Figure 4.15. Schematic view of potential feedback mechanisms in response to ice-shelf retreat/collapse. *Black* and *gray arrows* indicate positive and negative feedback, respectively. Not all potential feedback mechanisms are shown. See text for details

Surface albedo changes from ≈ 0.9 for snow-covered ice-shelves to ≈ 0.05 for open water. Thus, a change in shelf-ice extent affects net radiation and the surface-energy budget, and the albedo-temperature feedback has the potential to amplify warming. Positive feedback between lower albedo and warmer temperature potentially enhances warming. Open water is warmer (around the freezing point of ocean water) than is an ice shelf. An increase in open water extent can lead to substantial thermal destabilization. This feedback increases near-surface air temperatures. In summary, positive feedback between conditions resulting from ice shelf retreat/collapse and warmer temperatures may cause further ice-shelf retreat.

Depending on wind speed and shelf-ice structure, shelf ice may either be aerodynamically smoother or rougher than open water. Lower aerodynamic roughness, for example, reduces skin friction, enhances wind speed, and modifies wind direction in the lower ABL. Wind-field modification affects the transfer of momentum, water and energy fluxes, and atmospheric stability.

References

Alo CA, Wang G (2010) Role of dynamic vegetation in regional climate predictions over western Africa. Clim Dyn 35:907–922

Arneth A, Harrison SP, Zaehle S, Tsigaridis K, Menon S, Bartlein PJ, Feichter J, Korhola A, Kulmala M, O'Donnell D, Schurgers G, Sorvari S, Vesala T (2010) Terrestrial biogeochemical feedbacks in the climate system. Nat Geosci 3:525–532

Barnett TP, Adam JC, Lettenmaier DP (2005) Potential impacts of a warming climate on water availability in snow-dominated regions. Nature 438:303–309

Bhatt US, Walker DA, Raynolds MK, Comiso JC, Epstein HE, Jia G, Gens R, Pinzon JE, Tucker CJ, Tweedie CE, Webber PJ (2010) Circumpolar Arctic tundra vegetation change is linked to sea-ice decline. Earth Interact 14:1–20

Bokwa A (2010) Effects of air pollution on precipitation in Kraków (Cracow), Poland in the years 1971–2005. Theor Appl Climatol 101:289–302

Bonan GB (2008) Forests and climate change: forcings, feedbacks, and the climate benefits of forests. Science 320:1444–1449

Bounoua L, Collatz GJ, Sellers PJ, Randall DA, Dazlich DA, Los SO, Berry JA, Fung I, Tucker CJ, Field CB, Jensen TG (1999) Interactions between vegetation and climate: radiative and physiological effects of doubled atmospheric CO_2. J Clim 12:309–324

Chen J, Avise J, Guenther A, Wiedinmyer C, Salathe E, Jackson RB, Lamb B (2009) Future land use and land cover influences on regional biogenic emissions and air quality in the United States. Atmos Environ 43:5771–5780

Chow W, Pope R, Martin C, Brazel A (2011) Observing and modeling the nocturnal park cool island of an arid city: horizontal and vertical impacts. Theor Appl Climatol 103:197–211

Civerolo K, Hogrefe C, Lynn B, Rosenthal J, Ku J-Y, Solecki W, Cox J, Small C, Rosenzweig C, Goldberg R, Knowlton K, Kinney P (2007) Estimating the effects of increased urbanization on surface meteorology and ozone concentrations in the New York City metropolitan region. Atmos Environ 41:1803–1818

Claussen M, Brovkin V, Ganopolski A (2001) Biogeophysical versus biogeochemical feedbacks of large-scale land cover change. Geophys Res Lett 28:1011–1014

Dai A (2011) Drought under global warming: a review. Wiley Interdiscip Rev Clim Change 2:45–65

Diffenbaugh NS (2009) Influence of modern land cover on the climate of the United States. Clim Dyn 33:945–958

Euskirchen ES, McGuire AD, Rupp TS, Chapin FSI, Walsh JE (2009) Projected changes in atmospheric heating due to changes in fire disturbance and the snow season in the western Arctic, 2003–2100. J Geophys Res 114:G04022. doi:10.1029/2009JG001095

Fehrenbach U, Scherer D, Parlow E (2001) Automated classification of planning objectives for the consideration of climate and air quality in urban and regional planning for the example of the region of Basel/Switzerland. Atmos Environ 35:5605–5615

Field CB, Lobell DB, Peters HA, Chiariello NR (2007) Feedbacks of terrestrial ecosystems to climate change. Annu Rev Environ Resour 32:1–29

Grenfell TC, Perovich DK (2004) Seasonal and spatial evolution of albedo in a snow-ice-land-ocean environment. J Geophys Res 109:C01001. doi:10.1029/2003jc001866

Grimmond CSB, Oke TR (1995) Comparison of heat fluxes from summertime observations in the suburbs of four North American cities. J Appl Meteorol 34:873–889

Hallgren WS, Pitman AJ (2000) The uncertainty in simulations by a Global Biome Model (BIOME3) to alternative parameter values. Glob Change Biol 6:483–495

Heald CL, Henze DK, Horowitz LW, Feddema J, Lamarque JF, Guenther A, Hess PG, Vitt F, Seinfeld JH, Goldstein AH, Fung I (2008) Predicted change in global secondary organic aerosol concentrations in response to future climate, emissions, and land use change. J Geophys Res 113:D05211. doi:10.1029/2007jd009092

Hunt A, Watkiss P (2011) Climate change impacts and adaptation in cities: a review of the literature. Clim Change 104:13–49

Huntington TG (2006) Evidence for intensification of the global water cycle: review and synthesis. J Hydrol 319:83–95

Jonko A, Hense A, Feddema J (2010) Effects of land cover change on the tropical circulation in a GCM. Clim Dyn 35:635–649

Kim H, Kim S, Dale BE (2009) Biofuels, land use change, and greenhouse gas emissions: some unexplored variables. Environ Sci Technol 43:961–967

Kimball BA, Kobayashi K, Bindi M (2002) Responses of agricultural crops to free-air CO_2 enrichment. Advances Agronomy. 77:293–368

Lebourgeois F, Pierrat J-C, Perez V, Piedallu C, Cecchini S, Ulrich E (2010) Simulating phenological shifts in French temperate forests under two climatic change scenarios and four driving global circulation models. Int J Biometeorol 54:563–581

Lee JY, Kim YP, Kang C-H (2011) Characteristics of the ambient particulate PAHs at Seoul, a mega city of Northeast Asia in comparison with the characteristics of a background site. Atmos Res 99:50–56

Li Z (2007) Investigations on the impacts of land-cover changes and/or increased CO_2 concentrations on four regional water cycles and their interaction with the global water cycle. PhD thesis, Department of Atmospheric Sciences, University of Alaska Fairbanks, p 329

Li Z, Mölders N (2008) Interaction of impacts of doubling CO_2 and changing regional land-cover on evaporation, precipitation, and runoff at global and regional scales. Int J Climatol 28:1653–1679

Mölders N, Kramm G (2007) Influence of wildfire induced land-cover changes on clouds and precipitation in Interior Alaska – a case study. Atmos Res 84:142–168

Molina MJ, Molina LT (2004) Megacities and atmospheric pollution. J Air Waste Manage Assoc 54:644–680

Oke TR (1995) Boundary layer climates. Routlegde, New York

Oleson KW, Bonan GB, Feddema J (2010) Effects of white roofs on urban temperature in a global climate model. Geophys Res Lett 37: L03701. doi:10.1029/2009gl042194

PaiMazumder D, Mölders N (2009) Theoretical assessment of uncertainty in regional averages due to network density and design. J Appl Meteorol Climatol 48:1643–1666

Parry ML, Rosenzweig C, Iglesias A, Livermore M, Fischer G (2004) Effects of climate change on global food production under SRES emissions and socio-economic scenarios. Glob Environ Change 14:53–67

Pielke RA (2002) Mesoscale meteorological modeling. Academic, New York

Pielke RA Sr, Adegoke J, Beltrán-Przekurat A, Hiemstra CA, Lin J, Nair US, Niyogi D, Nobis TE (2007) An overview of regional land-use and land-cover impacts on rainfall. Tellus B 59:587–601

Pope ICA, Dockery DW, Schwartz J (1995) Review of epidemiological evidence of health effects of particulate air pollution. Inhal Toxicol 7:1–18

Ramos da Silva R, Werth D, Avissar R (2008) Regional impacts of future land-cover changes on the Amazon basin wet-season climate. J Clim 21:1153–1170
Reijnders L, Huijbregts MAJ (2008) Biogenic greenhouse gas emissions linked to the life cycles of biodiesel derived from European rapeseed and Brazilian soybeans. J Clean Prod 16:1943–1948
Sánchez E, Gaertner M, Gallardo C, Padorno E, Arribas A, Castro M (2007) Impacts of a change in vegetation description on simulated European summer present-day and future climates. Clim Dyn 29:319–332
Schurgers G, Mikolajewicz U, Gröger M, Maier-Reimer E, Vizcaíno M, Winguth A (2008) Long-term effects of biogeophysical and biogeochemical interactions between terrestrial biosphere and climate under anthropogenic climate change. Glob Planet Change 64:26–37
Searchinger T, Heimlich R, Houghton RA, Dong F, Elobeid A, Fabiosa J, Tokgoz S, Hayes D, Yu T-H (2008) Use of U.S. croplands for biofuels increases greenhouse gases through emissions from land-use change. Science 319:1238–1240
Sivakumar MVK (2007) Interactions between climate and desertification. Agric For Meteorol 142:143–155
Soja AJ, Tchebakova NM, French NHF, Flannigan MD, Shugart HH, Stocks BJ, Sukhinin AI, Parfenova EI, Chapin Iii FS, Stackhouse JPW (2007) Climate-induced boreal forest change: predictions versus current observations. Glob Planet Change 56:274–296
SRES (2000) Special report on emissions scenarios, working group III, Intergovernmental Panel on Climate Change (IPCC). Cambridge University Press, Cambridge
Taha H, Akbari H, Rosenfeld A (1991) Heat island and oasis effects of vegetative canopies: micro-meteorological field-measurements. Theor Appl Climatol 44:123–138
Verstraete MM, Scholes RJ, Smith MS (2009) Climate and desertification: looking at an old problem through new lenses. Front Ecol Environ 7:421–428
Walsh JE, Lynch A, Chapman W, Musgrave D (1993) A regional model for studies of atmosphere-ice-ocean interaction in the western Arctic. Meteorol Atmos Phys 51:179–194
Watson AJ, Lovelock JE (1983) Biological homeostasis of the global environment: the parable of Daisyworld. Tellus B 35B:284–289

Chapter 5

Conclusions

The previous chapters illustrated the substantial impacts of land-cover on the overlying atmosphere. Given these impacts, we cannot assume the atmosphere as constant factor that only changes in response to climate changes. This fact affects a wide range of applications from responsible environmental decisions to climate-impact assessment on water availability, food production, and air quality. Policymakers must be made aware that land-cover changes (LCC) influence at least local weather, climate, and atmospheric conditions. Before taking any action or making any decision, the unintended consequences thereof have to be assessed. These consequences may be adverse to the environmental problem to be solved, even engrave the problem or lead to other – even more severe – problems (Verstraete et al. 2009).

Holding domestic industries responsible for their environmental changes caused during the part of the production process they control may be hard to establish, economically unfeasible, and/or disadvantageous. The difficulty of enforcing such measures even increases in the case of greenhouse gas (GHG) emissions related to LCC. The GHG emissions caused by LCC are difficult to assess as they depend strongly on land management practices (fertilization, irrigation, crops rotation, tillage, etc.), type of LCC and duration of land-use (Kim et al. 2009). In some cases, taxing for LCC-induced GHG emissions could even become unethical when it would hinder to convert suitable land for urgently needed food production.

Assessment of LCC impacts on future water and food resources is difficult due to the complexity of the interactions of responses to LCC

N. Mölders, *Land-Use and Land-Cover Changes*, Atmospheric and Oceanographic Sciences Library 44, DOI 10.1007/978-94-007-1527-1_5,

among each other and climate change. In addition, the high spatial and temporal variability of precipitation and lack of accurate projections of LCC, GHG, and other emissions complicate the assessment.

The findings discussed in Chaps. 3 and 4 indicate LCC impacts on weather, climate, and the atmospheric composition at various scales and suggest feedback to further LCC. Atmospheric impacts of LCC can cause similar changes in regional temperature and precipitation like the changes in response to GHG. Thus, we have to conclude that LCC may have contributed to a certain degree to recent climate changes. This means that LCC are an important climate forcing that requires high research priority. Future modeling and observational studies should focus on quantifying the contribution of LCC to climate change.

5.1 Modeling and Observations

Modeling studies that assumed large, but realistic-size LCC indicate LCC-induced differences in 2 m air temperatures, surface temperatures, precipitation, and cloudiness larger than the accuracy of measurements. Thus, the impact of LCC could be detected if stations existed at the sites that experienced LCC. To obtain more observational evidence of LCC impacts on the atmosphere, existing long-term monitoring data in regions with extended LCC should be examined for changes related to documented LCC in the surroundings of sites. In regions of expected future LCC, installation of climate stations may help to quantify the impact of LCC on local weather and climate and to perform the overdue evaluation of LCC simulations. In LCC modeling, current common practice is to evaluate the model for current land-cover and atmospheric conditions and assume that the model performs similarly well with future land-cover conditions.

To document the actual role of the changing landscapes on weather, climate, and air composition, long-term monitoring sites should be established in areas of expected future LCC (e.g., deforestation) or land-use change (e.g., irrigation or expansion thereof). The monitoring efforts should focus on slow variables and their thresholds.

Any monitoring needs to cover the diversity of relevant land-cover. Since forest and grassland are still under-monitored, new sites have to be installed in these land-covers first. Both anthropogenic and natural LCC need to be recorded to permit associating them with climate changes (Verstraete et al. 2009). The latter helps to quantify the impact of LCC on weather and climate.

Such quantification also requires research on to which extent plants and/or ecosystems adapt their functional behavior to altered climate conditions and atmospheric composition and how biogeophysical and biogeochemical changes affect ecosystem functions (e.g., decomposition, nitrogen, and carbon cycling). These investigations require hierarchies of laboratory experiments that examine the ecosystem-function responses to various combinations of potential environmental changes. Improved ecosystem functions mean improved representation of future LCC, future emissions, energy and water needs in model studies, and reduction of uncertainty in the simulated resultant modified climate.

An important aspect of the value of LCC studies is the comparability of LCC impacts among studies performed for same LCC, but with different LCC extent and/or in different regions. To assess desertification and monitor dryland degradation, for instance, uniform criteria/standards of description and monitoring protocols have to be developed and adopted worldwide.

Since many LCC enhance the heterogeneity of the landscape, we need to develop new theories to interpret measurements with restricted or no fetch conditions or measurements in urban areas where surface characteristics vary too strongly as that a real "representative" site can be chosen. Such theories then may result in improved new parameterizations of the land-atmosphere interactions.

Earth observation from space is a cost-effective technology for monitoring LCC, and their impacts especially in remote, hard-to-reach regions. GOES-R, for instance, will have an increased resolution in many channels as compared to its predecessors; additional channels permit assessment of the chemical distribution of species. The integration of remote sensing data may improve our understanding of the significance of LCC-induced feedbacks. Better understanding of the feedbacks between climate and desertification, for instance, requires identifying the sources and sinks of trace gases and aerosols in drylands (Sivakumar 2007).

Satellite data could also help to assess and identify suitable areas for special field campaigns. Satellite-based passive microwave data may serve to detect changes in polynya, ice-shelves, and the extent and interannual variability of sea-ice. This data in combination with remotely sensed LCC can be used to analyze and evaluate ice-vegetation-atmosphere feedbacks.

Efforts have to be made to advance the development of (earth) system models that can describe the various interactions between biogeophysical and biogeochemical processes. Advances in LCC-impact and climate modeling require improved soil and vegetation parameters, root-depth specification, vegetation-distribution data and soil type profiles. We have to align monitoring and data collection with the modeling efforts to ensure that all of the needed soil and vegetation information is available for the models. Instead of classifying a region as covered by a certain soil type (e.g., clay loam), the data inventory should provide soil parameters (e.g., soil density, thermal and hydraulic conductivity, heat capacity, albedo, emissivity) as a 3D function in space. Similar applies for vegetation type.

Advances in earth-system modeling require to coupled models of different type. Prior to coupling models, it has to be evaluated at which scales which processes may interact and require the coupling. Such investigations will be the basis to decide whether to choose loosely coupled modeling techniques, one-way coupling or two-way-coupling. One-way coupling means that the driving model just provides the input for the other model without consideration of feedbacks. Two-way-coupling means that the results of the driven model feed back to the driving model and may cause changes in the state conditions and/or fluxes there. Besides the degree of coupling, one has to examine at which temporal intervals to induce the coupling. Thus, the time scales at which changes become important for other processes have to be determined. Any coupling requires developing routines to handle the required data exchange. Ideally, the system permits exchanging models addressing the same processes for sensitivity testing.

In any environmental decisions related to LCC, the feedbacks between the altered land surface and the atmosphere have to be considered on all spatial scales to assess their potential impacts on climate and air quality. Thus, future work has to focus on scaling issues (from local to regional to global scale) as the scale affects ecosystem functions. This research is especially urgent given the rapid growth of the number of megacities that are local LCC, but may be very sensitive to large-scale climate change.

Today, the knowledge on how spatially fine-scale LCC affect the large scale over longer time (e.g., seasonally) is very limited. The potentially far-reaching impacts of local LCC in a warmer climate (Li and Mölders 2008) require further investigations on the influence of land-cover on the climate system and the interaction between both.

The model grid increments determine which kind of land-atmosphere interactions can be resolved or must be parameterized and hence can provide atmospheric impacts. Mesoscale scale models can resolve processes that in large-scale models are of subgrid scale and have to be parameterized in large-scale models. Model-domain size, grid spacing, parameterizations (e.g., cloud microphysical schemes), and parameter choice affect significantly the distribution of simulated precipitation, chemical species, wet and dry deposition. Investigations must focus on consistency of responses at various scales and through the scales. Herein, the impacts of landscape pattern and heterogeneity on precipitation require systematic investigation. The effects of natural spatial variability and topography and climate variability have to be separated from changes due to anthropogenic LCC.

Despite the various LCC studies provide similar signals of LCC impacts, the LCC impacts on the atmosphere found vary largely among the various studies. This large variability requires uncertainty analysis on the influence of LCC. To increase the confidence in and assess the uncertainty of LCC simulations, simulations should be performed with different models that all assume the same LCC. Doing so requires guaranteeing comparable land-cover representations among participating models as differences due to land-cover representations may be of similar magnitude than those due to LCC (Pielke et al. 2007).

The knowledge on how concurrent LCC impacts interact with each other under different climate conditions is very limited. Future studies should focus on how the nonlinearity and significance of LCC found for current climate behaves on the long-term (e.g., vegetation season, decadal, or climate scale). If the LCC responses are climate-dependent, sophisticated biome models have to be developed that permit to include natural LCC in response to a changing climate. Social behavior models have to be run inline in the climate models to simulate economic responses (e.g., more land used for winter wheat) to a changing climate.

Coupling models inline ensures data consistency and facilitates to guarantee consistency of treatment. Offline simulations of LCC impact studies (e.g. impact on harvest, air quality, water management) have the advantage that more studies can be made, as the meteorology has not necessarily to be recalculated again for a new sensitivity study. However, meteorological data are not available for every time step. Interpolation and inconsistent treatment of processes by uncoupled models may lead to large errors (e.g., Mölders et al. 1994). For various aspects of the (earth) system, the strength of coupling must be examined. For systems that are only very loosely coupled, offline simulations may be a useful shortcut and tool for huge numbers of sensitivity studies.

Studies on LCC impacts also bear uncertainty from the unknown development of land management practices. New land management practices may change tillage, rotation of crops, fertilization, and irrigation or soil density with consequences for infiltration. Sensitivity studies have to be performed to assess land management-related uncertainty in LCC impacts on weather, climate, and atmospheric composition.

The efforts to reduce emission-related climate change or to improve air quality can themselves affect local and regional climate via direct and indirect aerosol effects. Thus, future climate projections must include LCC, aerosols, and trace gases to consider interactions between the altered energy, water, and trace gas cycles. This means studies assessing future or potential long-term changes on local and regional climate should consider LCC and GHG forcing concurrently.

Various processes are still underrepresented in GCMs, mesoscale, or air-quality models due to computer-time limitations. The anticipated growth and number of megacities requires assessing urbanization impacts, including urban heat island (UHI) and air-quality issues. Thus, urban canopy models should be run inline in climate models to assess future quality of life and to be able to use these models to find means to mitigate the UHI stress and air pollution. Running the urban canopy and climate models in a fully coupled mode permits to consider the feedbacks between the local scale and larger scales.

Studies show that CO_2-induced and climate-induced changes in vegetation structure can influence hydrological processes and climate at similar or higher magnitudes than the radiative and physiological effects (e.g., Alo and Wang 2010). Therefore, including vegetation feedbacks in future climate projections is an urgent need. Future scenarios should include relevant economic and environmental aspects that alter land-surface conditions, and land-cover should change accordingly during the simulation. Such aspects could be, for instance, reduced irrigation in regions that experience either an increase or strong stresses on water availability.

5.2 Future Assessment

Precipitation anomalies in response to urban effects (UHI, enhanced buoyancy, reduced evapotranspiration, altered cloud condensation, and ice nuclei distributions) can yield ecological and societal consequences of both signs. Plans for ameliorating negative consequences might include the development of improved water-management strategies, improved guidelines for home construction, and/or recommendations for optimal location of industrial or commercial areas. In such planning processes, results from atmospheric models in combination with statistical methods (e.g., analysis of variance) can help in identifying which of the different possible urban effects actually causes the precipitation variance. Such an identification process avoids arbitrary and possibly ineffectual changes and can help policymakers to make cost efficient decisions, and implement effective policies.

The changes in technology and energy supply bear great uncertainties for any future assessments. These changes namely may alter the atmospheric conditions via emissions, and land-cover for food and biofuel production. Some scientists warn that the assumptions of the Intergovernmental Panel on Climate Change (IPCC) Special Report on Emissions Scenarios (SRES) may not provide the full picture of future challenges (Pielke et al. 2008). According to them, the SRES-assumed rapid decline in energy intensity of greater than 1% per year over a century would require probably unachievable advances in energy efficiency; energy demands are more likely to increase than decrease given the increase in population and the economic transition in developing countries (Pielke et al. 2008). Maintaining a megacity, for instance, requires huge amounts of energy for pumping water into high buildings, cleaning wastewater, providing air conditioning or heating, and transportation of people and/or food supply.

Economic or social demands often conflict with environmental interests. A huge challenge for policymakers is to tackle these conflicts of interest. Typically, environmental policies related to biodiversity or water quality gain high public acceptance. However, policymakers usually neglect local climate impacts in urban and/or regional planning due to the lack of appropriate tools for assessing the impact of LCC on local climate (Fehrenbach et al. 2001).

In a changing world, big issues are that LCC and climate changes affect the welfare of states differently, and that the LCC, air pollution, and GHG impacts do not know international borders. This fact requires worldwide efforts and international collaboration to keep impacts as small as possible while at the same time addressing the energy, water, and food demands of an increasing world population.

References

Alo CA, Wang G (2010) Role of dynamic vegetation in regional climate predictions over western Africa. Clim Dyn 35:907–922

Fehrenbach U, Scherer D, Parlow E (2001) Automated classification of planning objectives for the consideration of climate and air quality in urban and regional planning for the example of the region of Basel/Switzerland. Atmos Environ 35:5605–5615

Kim H, Kim S, Dale BE (2009) Biofuels, land use change, and greenhouse gas emissions: some unexplored variables. Environ Sci Technol 43:961–967

Li Z, Mölders N (2008) Interaction of impacts of doubling CO_2 and changing regional land-cover on evaporation, precipitation, and runoff at global and regional scales. Int J Climatol 28:1653–1679

Mölders N, Hass H, Jakobs HJ, Laube M, Ebel A (1994) Some effects of different cloud parameterizations in a mesoscale model and a chemistry transport model. J Appl Meteor 33:527–545

Pielke RAS, Adegoke J, Beltrán-Przekurat A, Hiemstra CA, Lin J, Nair US, Niyogi D, Nobis TE (2007) An overview of regional land-use and land-cover impacts on rainfall. Tellus B 59:587–601

Pielke R, Wigley T, Green C (2008) Dangerous assumptions. Nature 452:531–532

Sivakumar MVK (2007) Interactions between climate and desertification. Agric For Meteorol 142:143–155

Verstraete MM, Scholes RJ, Smith MS (2009) Climate and desertification: looking at an old problem through new lenses. Front Ecol Environ 7:421–428

List of Abbreviations

ABL	Atmospheric boundary layer
AGL	Above ground level
AQF	Air quality forecast
Ar	Argon
ARME	Amazon region micrometeorology experiment
ASL	Atmospheric surface layer
AST	Alaska standard time
AVHRR	Advanced very high resolution radiometer
BOREAS	BOReal ecosystem-atmosphere study
CAM3	Community atmospheric model version 3
CAPE	Convective available potential energy
CASES	Cooperative atmosphere surface exchange study
CCN	Cloud condensation nuclei
CH_3	Methyl radical
CH_3O	Methoxyl radical
CH_3O_2	Methylperoxyl radical
CH_4	Methane
CIN	Convective inhibition
CLM	Community land model
CO	Carbon monoxide
CODE	California ozone deposition experiment
CO_2	Carbon dioxide
DJF	December-January-February
EFEDA	European field experiment in a desertification-threatened area
ENSO	El Niño/southern oscillation
EU	European union
Fe^{3+}	Iron
FIFE	First ISLSCP field experiment

N. Mölders, *Land-Use and Land-Cover Changes*, Atmospheric and Oceanographic Sciences Library 44, DOI 10.1007/978-94-007-1527-1,

GCM	Global circulation model
GEWEX	Global energy and water cycle experiment
GHG	Greenhouse gases
GIS	Geographic information system
H	Hydrogen atoms
HAPEX	Hydrological and atmospheric pilot experiment
HCHO	Formaldehyde
HNO_3	Nitric acid
HO_2	Hydroperoxyl radicals
HRV	High resolution visible
HSO_3	Bisulfite
HSO_3^-	Bisulfite ions
HSO_4^-	Bisulfate ions
HTSVS	Hydro-thermodynamic soil vegetation scheme
H_2CO	Formaldehyde
H_2O	Hydrogen oxide
H_2O_2	Hydrogen peroxide
H_2SO_3	Sulfurous acid
H_2SO_4	Sulfuric acid
ICA	Independent component analysis
IN	Ice nuclei
IPCC	Intergovernmental Panel on Climate Change
ISLSCP	International satellite land surface climatology project
JJA	June-July-August
LAI	Leaf area index
LC	Land conversion
LCC	Land-cover changes
LCL	Lifting condensation level
LFC	Level of free convection
LOTREX	Longitudinal land-surface transverse experiment
MCS	Mesoscale convective system
MEGAN	Model of emissions of gases and aerosols
METROMEX	Metropolitan meteorological experiment
MM5	Meteorological model generation 5
Mn^{2+}	Manganese
MOBILHY	Modélisation du bilan hydrique
MODIS	Moderate resolution imaging spectroradiometer
MRF	Markov random field models
MSS	Multi-spectral scanner
NAAQS	National ambient air quality standards
NCMC	Non-classical mesoscale circulations
NDVI	Normalized difference vegetation index

NH_3	Ammonia
NIR	Near-infrared reflectance
NO	Nitric oxide
NO_2	Nitrogen dioxide
NO_3	Nitrate radical
NO_x	$=NO + NO_2$
NWP	Numerical weather prediction
N_2	Nitrogen
N_2O	Nitrous oxide
OH	Hydroxyl radicals
OSULSM	Oregon State University land surface model
O^{1D}	Exited oxygen atom
O_2	Oxygen
O_3	Ozone
PAH	Polycyclic aromatic hydrocarbons
PAR	Photosynthetic active radiation
PDO	Pacific decadal oscillation
PFT	Plant functional types
PM	Particulate matter
POES	Polar operational environmental satellite
QOL	Quality of life
RAMS	Regional atmospheric modeling system
RBLE	Rondonia boundary layer experiment
RCH(OH)O	Hydroxyalkoxy radical
SAR	Synthetic aperture radar
SL	Sea level
SMA	Spectral mixture analysis
SOA	Secondary organic aerosols
SO_2	Sulfur dioxide
${SO_3}^{2-}$	Sulfite ions
${SO_4}^{2-}$	Bisulfate ions
SPOT	Satellite pour l'Observation de la Terre
SRES	Special report on emission scenarios
SST	Sea-surface temperature
SVM	Support vector machine
SWI	Summer warmth index
TM	Thematic mapper
UHI	Urban heat island
VIS	Visible spectral range
VOC	Volatile organic compounds
WRF	Weather research and forecast model

Index

N. Mölders, *Land-Use and Land-Cover Changes*, Atmospheric and Oceanographic Sciences Library 44, DOI 10.1007/978-94-007-1527-1,

9789400715264